THE PERIODIC KINGDOM

science masters

Already published in the SCIENCE MASTERS series:

JOHN D. BARROW
The Origin of the Universe

PAUL DAVIES
The Last Three Minutes

RICHARD DAWKINS
River Out of Eden

RICHARD LEAKEY
The Origin of Humankind

IAN STEWART
Nature's Numbers

Future contributors to the series include:

MARY CATHERINE BATESON
COLIN BLAKEMORE
WILLIAM H. CALVIN
DANIEL C. DENNETT
JARED DIAMOND
MURRAY GELL-MAN
STEPHEN JAY GOULD
SUSAN GREENFIELD
DANIEL HILLIS
LYNN MARGULIS
MARVIN MINSKY
STEVEN PINKER
STEPHEN H. SCHNEIDER
GEORGE SMOOT
ROBERT A. WEINBERG
GEORGE C. WILLIAMS

PETER ATKINS

THE PERIODIC KINGDOM

A Journey into the Land of the Chemical Elements

WEIDENFELD & NICOLSON · LONDON

First published 1995

SCIENCE MASTERS are published in association with
Basic Books, a division of HarperCollins Publishers, New York.

Printed in Great Britain by
Clays Ltd, St Ives plc

Weidenfeld & Nicolson
The Orion Publishing Group
Orion House
5 Upper Saint Martin's Lane
London WC2H 9EA

British Library Cataloguing-in-Publication Data
A catalogue record for this book is available
from the British Library

ISBN 0 297 81641 1

CONTENTS

PREFACE

I have always been struck by the opening of Somerset Maugham's *The Vessel of Wrath*. The author sits in his study, leafing through the *Yangtse Kiang Pilot*, and in his mind's eye the tide tables and navigation directions gradually take on a sense of reality. In his imagination the contours and tables give way to a richer version, as he becomes aware of trees, roofs, and finally the people who are the subjects of his tale. So I would like you to accompany me on a journey of imagination through the austere navigation chart of chemistry, the periodic table of the elements. But in our mind's eye we shall see it as a country—the Periodic Kingdom—populated, as we shall see when we descend to its surface, by personalities. We shall fly through the landscape of the kingdom, see its rolling hills, its mountain ranges, its gorges, and its plains. We shall land, and walk among its broad meadows and across its hills. We shall even burrow beneath the surface, and discover that there is a hidden structure, a mechanism, that controls and governs the kingdom. For this is a rational place.

The periodic table is arguably the most important concept in chemistry, both in principle and in practice. It is the everyday support for students, it suggests new avenues of research to professionals, and it provides a succinct organization of the whole of chemistry. It is a remarkable demonstration of the fact that the chemical elements are not a random clutter of entities but instead display trends and lie together in families. An awareness of the periodic table is essential to anyone who wishes to disentangle the world and see how it is built up from the fundamental building blocks of chemistry, the chemical elements. Anyone who seeks to be familiar with a scientist's-eye view of the world must be aware of the general form of the periodic table, for it is a part of scientific culture.

I have presented the periodic table as a kind of travel guide to an imaginary country, of which the elements are the various regions. This kingdom has a geography: the elements lie in particular juxtaposition to one another, and they are used to produce goods, much as a prairie produces wheat and a lake produces fish. It also has a history. Indeed, it has three kinds of history: the elements were discovered much as the lands of the world were discovered; the kingdom was mapped, just as the world was mapped, and the relative positions of the elements came to take on a great significance; and the elements have their own cosmic history, which can be traced back to the stars.

The Periodic Kingdom also has an administration, for the elements have properties governed by laws that control their behavior and determine the alliances they form. This administration is to be found in the properties of atoms, and of the electrons and nuclei that constitute atoms.

I have not presumed any prior knowledge of chemistry. All I ask is that you use your imagination to interpret the geographical analogies in terms of concrete entities. We shall fly through the landscape together, and land when it suits us. In this way we shall discover a rich kingdom, of which our actual world is a manifestation.

I would like to thank Jerry Lyons, who edited this book, for a number of helpful suggestions, and Sara Lippincott, who as copyeditor helped admirably to elucidate what I wanted to say.

Oxford, January 1995

THE PERIODIC KINGDOM

PART I

GEOGRAPHY

CHAPTER 1

THE TERRAIN

Welcome to the Periodic Kingdom. This is a land of the imagination, but it is closer to reality than it appears to be. This is the kingdom of the chemical elements, the substances from which everything tangible is made. It is not an extensive country, for it consists of only a hundred or so regions (as we shall often term the elements), yet it accounts for everything material in our actual world. From the hundred elements that are at the center of our story, all planets, rocks, vegetation, and animals are made. These elements are the basis of the air, the oceans, and the Earth itself. We stand on the elements, we eat the elements, we *are* the elements. Because our brains are made up of elements, even our opinions are, in a sense, properties of the elements and hence inhabitants of the kingdom.

The kingdom is not an amorphous jumble of regions, but a closely organized state in which the character of one region is close to that of its neighbor. There are few sharp boundaries. Rather, the landscape is largely characterized by transitions: savannah blends into gentle valleys, which gradually deepen into almost fathomless gorges; hills gradually rise from plains to become towering mountains.

These are the images, the analogies, to keep in mind as we travel through the kingdom. The principle to keep in mind is that not only is the material world built from a hundred or so elements but these elements also form a pattern.

At this stage, all we shall aim to achieve is an appreciation of the pattern, a familiarity with the countryside of the kingdom. From overhead we see that it stretches in a great vista, from hydrogen to beyond distant uranium. There are known regions beyond uranium, but in the very far distance lies an unexplored horizon awaiting a new Columbus. Closer to us is more familiar territory—regions such as carbon and oxygen, nitrogen and phosphorus, chlorine and iodine. By the end of this initial survey, we shall be familiar with many more regions, and know them as the equivalent of desert, swamp, lake, or pasture.

Even from this height, far above the country, we can see broad features of the landscape (fig. 1). There are the glittering, lustrous regions made up of metals and lying together in what we shall call the Western Desert. This desert is broadly uniform, but there is a subtlety of shades, indicating a variety of characteristics. Here and there are gentle splashes of color, such as the familiar glint of gold and the blush of copper. How remarkable it is that these desert lands make up so much of the kingdom (about 86 of the 109 known elements), yet the kingdom supplies such luxuriance in the real world! This abundance suggests that the desert's barrenness is an illusion, and that when we are closer we shall see that it is rich in the equivalent of minerals and that its bleak landscape possesses a wide range of physical and chemical properties. But closer inspection is for later. Now we are still far above the surface of the kingdom, and the desert's variety is not readily apparent.

To the east, the landscape varies markedly, even from

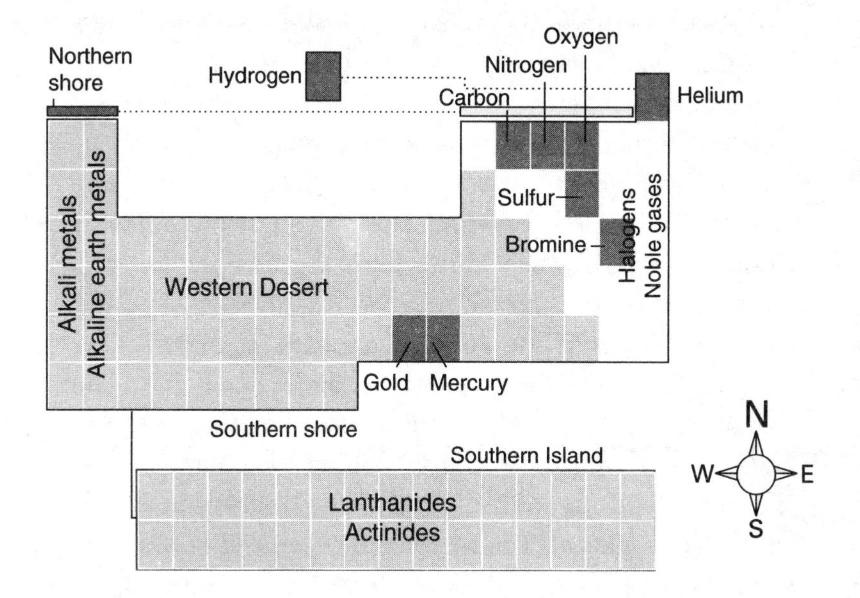

FIG. 1.

The general layout of the Periodic Kingdom, with some of its characteristic regions labeled. The Western Desert and the Southern Island comprise the metallic elements; all other elements are nonmetals.

....

this altitude. Here lie the softer regions of the kingdom, and a lake can be seen. But it is immediately clear that this is no ordinary kingdom, for the lake is not the limpid gray or blue of Earthly lakes but a striking deep red, verging on brown. This region is known as bromine, and it is one of only two lakes in this peculiar land. The other lake is on the eastern edge of the Western Desert, and is quite different in appearance, with a harshly metallic, silvery sheen. This is mercury, a liquid lake amidst the stone.

In these eastern lands, there is great variety of form and color, and the variety increases as we approach the eastern coast. The Western Desert has gradually modulated into regions that are seemingly metallic but have a softened personality, an ambiguity of character. Such elements include silicon and arsenic, and less familiar regions of the kingdom such as polonium and tellurium. Seemingly, the land is becoming chemically fertile, but in this unfamiliar countryside, the reality may differ from the first impression. Most striking from this altitude, we see the colors of the landscape. There is the familiar vivid splash of sulfurous yellow—the brimstone of yesteryear and hell. Its neighbor selenium varies, as though with the seasons, from metallic gray in one form to ruby red in another. How can a single substance take on such different hues? Selenium is not alone in this respect. The common element carbon also has remarkable variety: its most familiar form is sooty black, but it can modulate to sparkling diamond, gray metallic graphite, and the orange-brown of a recently discovered crystalline form called fullerite. We need to be aware of these various forms of a given element—forms we shall allegorically term its seasons, but which are actually termed its allotropes—because ignorance of them can lead to confusion. When we descend to the surface of the kingdom, we may find the elements in any one of their seasons.

The colors of the landscape become even more varied near the eastern coastline. Most striking of all are the *halogens*, a close-knit family of regions that includes the red lake of bromine. From this altitude, we see the gradation of their colors, from almost colorless fluorine in the far north, deepening through yellow-green chlorine to the neighboring ruddy bromine. South of bromine lies the lustrous purplish-black iodine, shimmering close to the southern shoreline. And south of iodine lies astatine, a region that has a name, but about which little else is known. Such ignorance is allied, as it so often is in the real world, with uselessness; if you have never heard of astatine, it is because there is apparently little point in knowing anything about it. It is one of the unexploited, unproductive, and only superficially investigated regions of the kingdom.

The variation of color in this zone is one of the more obvious aspects of the gradation of properties of the elements here. This part of the kingdom attracts the eye with its pleasing hues, and we are led to think there might be subtle trends needing closer inspection, even excavation, to discern. That is in fact the case, and when we travel through the kingdom on foot, we must be alert for modulations that do not show themselves so plainly. A part of the pleasure of chemistry—the name for the laws governing the Periodic Kingdom—rests in the discovery of deep underlying rhythms that span the kingdom's regions, uniting them into families.

One of these trends is essentially invisible to the naked eye. Just as we cannot see the air we breathe, some regions of the Periodic Kingdom appear to be devoid of substance. These seemingly empty regions lie on the eastern and northeastern coasts, landscapes even less fertile in appearance than the Western Desert. But the northeastern coast is far from useless, for here lies oxygen, that almost universal harbinger of life. Oxygen is so vital to organic life on Earth today that where it is not available it must be supplied. We carry it beneath the sea in tanks; we conveyed it to the moon. We pump it into moribund bodies to help them keep their grip on life. We squirt it by the ton into engines to enable them to burn their fuel. Oxygen is the essence of animation, and animation and purposeful locomotion normally cease in its absence. Such is the hidden potency of this seemingly substance-free region on the northern fringe of the kingdom.

Invisible potency is not confined to oxygen. Its western neighbor, nitrogen, is also apparently without substance but essential to life: much biological and industrial chemical activity is focused on its capture from the atmosphere, where it is found in extraordinary abundance. The capture of nitrogen is called its "fixation," and nitrogen fixation is a process of vital importance on Earth—as vital as photosynthesis, the fixation of carbon from atmospheric carbon dioxide. Nitrogen fixation had to be achieved long before humans appeared on Earth, because proteins are built from it, and proteins are essential to all varieties of organic life. Even the handing down of heritable information from generation to generation relies on nitrogen, because it is a component of deoxyribonucleic acid, or DNA, the stuff of genes. Without the apparently blank region of nitrogen, life would terminate: not only would there be no heritability but there would be no activity of any sort—for the gearwheels of life, the proteins, would not exist.

The regions of the eastern coast are a different matter. They, too, are gaseous, but they are largely inactive. They have gone by various names since they were first discovered by chemical explorers at the end of the nineteenth century. They were originally called the rare gases, for they were believed to be scarce. That is certainly true of some of them but not of all of them: one of these gases, argon, is more abundant in the Earth's atmosphere than carbon dioxide. Helium, while rare in our atmosphere, is far from rare in the universe, where it makes up 25 percent of all there is and is second in abundance only to hydrogen. And radon, far south on the eastern coast, is dangerously abundant in parts of the Earth where natural radioactivity occurs. "Rare" is hardly a suitable way to characterize these plentiful gases, and the name has been abandoned. They were once also known as the inert gases;

one reason this coastal fringe remained undiscovered for so long was that these elements were not found in combination. To the early chemical explorers, who made their discoveries by examining combinations of regions, the lands of the coastline remained invisible and even unsuspected. However, more recent attempts to coax them into combination have been successful, and the infertility of these regions has been overcome. That is not to say the coastal desert now blooms, but there is an occasional glimmer of fertility, the chemical equivalent of a blade of grass. So, gone too is the justification for "inert." These regions are now known collectively as the *noble gases*, a name intended to imply a kind of chemical aloofness rather than a rigorous chastity.

In summary, the general disposition of the land is one of metals in the west, giving way, as you travel eastward, to a varied landscape of nonmetals, which terminates in largely inert elements at the eastern shoreline. To the south of the mainland, there is an offshore island, which we shall call the Southern Island. It consists entirely of metals of subtly modulated personality. North of the mainland, situated rather like Iceland off the northwestern edge of Europe, lies a single, isolated region—hydrogen. This simple but gifted element is an essential outpost of the kingdom, for despite its simplicity it is rich in chemical personality. It is also the most abundant element in the universe and the fuel of the stars.

CHAPTER 2

THE PRODUCTS OF THE REGIONS

We have seen in the cases of oxygen and nitrogen that invisibility does not mean uselessness. Likewise, the regions of the almost uniform metallic desert are rich sources of fecundity when appropriately deployed. To survey the entire kingdom for its products would be inappropriate at this stage, for we need to know more about the kingdom's institutions before we can appreciate the products fully. But an introductory survey is appropriate.

The metallic elements are of crucial importance, both in the natural landscape of the real world and in the artifacts contrived by science and industry. For instance, one region of the Western Desert is iron, the element that helped to lift humanity out of the Stone Age and impel it toward and through the Industrial Revolution. In the course of that revolution, this region boomed. By forming alliances with a number of its neighbors, such as cobalt, nickel, vanadium, and manganese, iron became steel, and steel is almost literally the foundation of modern society. The fact that iron is able to form alliances so readily with its neighbors is a point not to go unnoticed in the kingdom. This feature indicates that there are deep currents of similarity beneath the surface of the kingdom's landscape, analogous to the cultural and economic ties that reinforce alliances between nations.

The regions near iron have also been particularly significant in the history of actual nations. A few steps to the east of iron lies copper, which—because it could be extracted so easily from its ores—was the first element to be used on our long journey out of the Stone Age, and among the first materials ever to be processed. Copper is moderately resistant to the unwanted chemical change we term corrosion. There are two everyday consequences of this resistance: one is the use of copper in water pipes—water being one of the most virulent of all chemicals—and the other is its use, in alliance with the nearby regions of bronze and nickel, in coinage. Here we see another of the kingdom's subterranean links. Close to copper lie silver and gold, which have long been used as metals of commerce, decoration, and coinage, partly because of their attractive appearance and their rarity but also because they, too, are resistant to corrosion; indeed, copper, silver, and gold are sometimes termed the coinage metals, in recognition of their role in society.

Generally speaking, the Western Desert was explored and exploited from east to west; that is, technology and industry made use of them in that order. Copper displaced stone to give us the Bronze Age. Then, as the explorers pressed westward, applying ever more vigorous means of discovery, they encountered the region of iron and used it to fabricate more effective weaponry. Under the pressures of natural selection, those societies that made use of such weapons—being better equipped for killing, subjugation, and survival—flourished. The strongest states enjoyed freedom from constant aggression, and this gave them time for scholarship; thus in due course explorers were able to penetrate into more distant western regions of the desert.

Here they discovered extraordinary riches. Deep in the Western Desert, in an isthmuslike zone running from zinc on the east to scandium on the west, they finally stumbled upon titanium, a remarkable prize indeed. Titanium has exactly the properties that a society bent on high technology needs if it is to take to the skies: this is a metal that is tough and resistant to corrosion, yet light, and it is typical of its part of the Western Desert. Titanium and its neighbors vanadium and molybdenum, in alliance with iron, form the durable steels that enable us to chop through stone and build on a massive scale. It is a remarkable feature of the Periodic Kingdom that the Isthmus, which links two rectangular blocks at the kingdom's western and eastern extremes, has provided so many of the workhorses of our society, and that its members so readily form alliances with one another.

The Western Rectangle is also part of the Western Desert, but its regions are somewhat more striking than those of the Isthmus. Here are the metals that are never found native (the technical term for their free form) in nature, because they are highly reactive. Most of these metals are too virulent even to touch. Here is a chemical feature of the kingdom which, although not as immediately obvious as the colors of the Eastern Rectangle, is nevertheless no less striking. To perceive it we need only witness a simple reaction: we need to watch what happens when it rains in this remote part of the kingdom.

In the far northwestern region of lithium, rain has little effect. Certainly the terrain simmers and bubbles with hydrogen gas as the rain hits the ground, but on the whole the reaction is quiet and the countryside is little disturbed. That is not so with sodium, which borders it on the south. Here rain and terrain are in violent conflict; the ground seethes and boils wherever a raindrop strikes. If rain is tolerable in the region of lithium and scarcely bearable in the region of sodium, rain in potassium, the region immediately to the south of sodium, is beyond belief. Now not only does the ground heave and boil, but it ignites and burns. So vigorous is the reaction between metal and rainwater here that the hydrogen catches fire; this region is uninhabitable in the rain. And still further south? Now rain is explosive. In the regions of rubidium and cesium, each raindrop is a bomb, riven by explosion on impact with the ground. The trend in this family of elements is of increasing chemical reactivity from north to south.

The virulence of these metals of the coast of the Western Desert—collectively known as the *alkali metals*—does not mean they are useless in nature and industry. Virulence, when controlled, has its uses. Sodium, for instance, is a component of table salt (sodium chloride), a substance so useful that governments have regulated its distribution. Sodium is an essential component of the activity of the nervous system and brain, and without it we elaborate organisms would be inanimate and largely functionless plants. Potassium, an alkali metal subtly different in character from sodium, is also an essential component of neuronal activity, and the cautious interplay between these two similar elements sustains thought and action, thereby animating the otherwise inanimate. Here again we see the potential of alliance between neighbors in the kingdom: a harmony of properties can be much richer in its consequences than properties that operate alone. Thought is the most subtle phenomenon of matter; appropriately, it stems, in part, from the interplay of two neighboring regions between which there is only a subtle difference in properties.

Just to the east of the alkali metals on this rectangular block of land lies another group of regions with closely related properties. This family of regions, comprising the ponderously named *alkaline earth metals*, includes calcium, which also seethes when it rains, quietly bubbling and forming hydrogen rather as lithium does. Calcium, though, is much more useful than lithium, and nature found uses for this region of the Western Desert long before human civilizations began using lithium (principally in nuclear bombs). Calcium is a component of nervous activity, like sodium and potassium, but it is also an element characteristic of construction and the maintenance of form. In nature it is a component of such materials as the bone (calcium phosphate) of endoskeletons and the shells (calcium carbonate) of exoskeletons. The accumulation of the discarded calcium carbonate of shellfish has ultimately contributed to the rigid endoskeletons of our own Earthly landscape. Limestone ridges are the legacy of marine life, and endure by virtue of the calcium they contain.

Nature's use of calcium has been replicated by human civilizations, which have quarried limestone and constructed buildings that have lasted thousands of years. The Romans manufactured concrete and mortar but did not realize that they were in effect mining calcium from the kingdom. Without this product of the region of the Western Desert, not only would there be no permanent structures in civilized societies but organisms would never have developed weapons of attack (no teeth or tusks) or protection in the form of shells.

Just to the north of calcium lies magnesium, a region with similar but, again, subtly different properties. Magnesium is less reactive than calcium: when it rains here, the terrain remains largely unchanged. But like calcium, magnesium can act as a backbone of sorts, and it is found along with calcium in the chalklike mineral called dolomite, of which the Dolomites of Austria and Italy are constructed. A particularly important product of the region of magnesium is an organic molecule called chlorophyll, which contains a single magnesium atom at its eye. Without chlorophyll, the world would be a damp warm rock instead of the softly green haven of life that we know, for chlorophyll holds its magnesium eye to the sun and captures the energy of sunlight, in the first step of photosynthesis. For reasons we shall explore, magnesium has exactly the right features to make this process possible. Had the kingdom lacked this element, chlorophyll's eye would have been blind, photosynthesis would not take place, and life as we know it would not exist.

At the southern end of this family are the metals strontium, barium, and radium. The kingdom's patterns are beginning to be established, and since patterns are the foundation of prediction we are able to predict that these regions will be much more reactive than those to the north. Indeed, they are too aggressive to their environment to be of much use, and nature has found no use for them. Nature's child, humanity, though, has put them to use. Radium is highly radioactive (a nuclear, not a chemical, property), and is used to kill unwanted proliferating cells. A radioactive form of strontium, strontium-90, is a component of nuclear fallout, and if it accumulates in place of calcium in bone it can kill cells that are needed for life and induce leukemia.

It is time to cross the Isthmus again, to the rectangular land in the East. The products of the Eastern Rectangle are rich beyond belief. The most interesting regions lie on the northern coast; two of them, oxygen and nitrogen, we have already visited. But no region of the northern coast—or of anywhere else in the kingdom, for that matter—is more fecund than carbon. Carbon is a particularly mediocre element, easygoing in the liaisons it forms. Unlike fluorine, which is only a few regions to the east of it, carbon is not a reactive prima donna. In chemistry as in life, this unpretentiousness has rewards, and in its mediocre way carbon has established itself as king of the Periodic Kingdom. Carbon, of course, is the element of organic compounds: the extraordinary and complex property we term "life" stems almost in its entirety from this black northern region of the kingdom.

Immediately to the south of carbon lies silicon. As is so often the case with neighbors, it is an uneasily ambiguous adjacency. Like carbon, but to a lesser degree, silicon is capable of forming some of the long-chain molecules needed in any process as complex as life, but it has not achieved a life of its own. It may be a sleeper in this regard, however. Carbon's principal products, living organisms, have struggled over a few billion years to establish mechanisms for the accumulation and dispersal of information (an austere distillation and definition of what we mean by "life"), and silicon has lain in wait. The recent alliance of the two regions, in which carbon-based organisms have developed the use of silicon-based artifacts for information technology, has resulted in the enslavement of silicon. However, such is the precocity of carbon's organisms that they are steadily developing silicon's latent powers, and one day silicon may well overturn the suzerainty of its northern neighbor and assume the dominant role. It certainly has long-term potential, for its metabolism and replication need not be as messy as carbon's. Here we may see one of the most subtle interplays of alliances anywhere in the kingdom, for silicon will not realize its potential without the burden of development being carried out by carbon.

One particular rhythm of the kingdom should be noticed while we are hovering in this zone. Carbon is a reasonably typical nonmetallic element, just as are its western and eastern neighbors, boron and nitrogen. However, if we fly directly south we will move into the fringe of the Western Desert, which juts diagonally—from northwest to southeast—into the Eastern Rectangle. In this jutting triangle lie the embers of the desert: familiar metallic elements like aluminum, tin, and lead and unfamiliar ones like gallium and thallium. Now the pattern to note is that of diagonality: frontiers and similarities form northwest to southeast diagonals across the kingdom.

Whereas the metals of the western coastline are so vigorous as to be dangerous in their reactivity, here on the desert's eastern annex the metals are much calmer—calm to the point of chemical indifference. Tin, for example, was once applied as a protective coating to steel cans, before tin's nearby northern neighbor aluminum displaced both skeleton (iron) and skin (tin) for the almost universal packaging of beverages, even those as corrosive as colas. Lead, tin's southern neighbor, had an even longer run through history on account of its chemical indifference. From the time of the Roman Empire to the present day, it has been used for the conduction of that corrosive liquid, water. It is rather a shame that lead is not completely inert in the presence of water, and that small amounts perfused into human brains can lead to mental debilitation and thence to the collapse of empires. Thus it can be with apparent allies—in this case, carbon in the distant north and deep southerly lead.

The distinction between the metals of the eastern and western fringes of the Western Desert—the former calm and contented, the latter virulent and violent—suggests that the Isthmus forms a transitional bridge between the two (fig. 2).

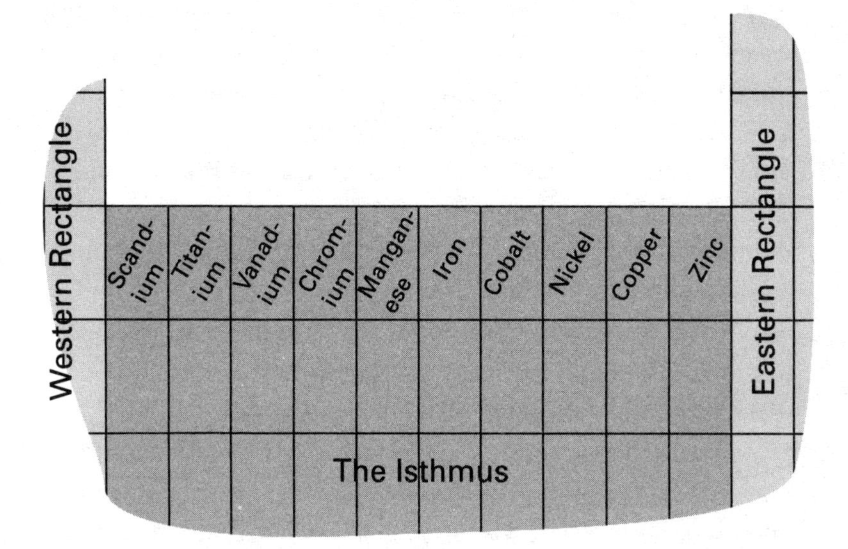

FIG. 2.

Some of the metals that make up the Isthmus, which joins the Western and Eastern Rectangles.

....

That transitional nature is apparent even in the Isthmus itself: scandium, in the west, is a violent metal, whereas copper, which is close to the eastern edge of the Isthmus, is calm. Indeed, the elements of the Isthmus are called the *transition metals.* There are certain technical reasons that exclude the eastern abutment of the Isthmus—zinc, cadmium, and mercury—from this appellation, but broadly speaking we can think of the Isthmus as representing a slope of declining chemical activity from west to east.

To return to the Eastern Rectangle: At the diagonal border of the Western Desert—the northwest-to-southeast line where the metals peter out and give way to the nonmetals—there is an ambiguous zone of elements known as *metalloids*, which possess metallic and nonmetallic characteristics. This diagonal zone includes familiar regions, like silicon and arsenic, and some regions perhaps less familiar, like antimony and polonium. It is to be noted that carbon abuts this zone and that silicon lies within it. It is probably no coincidence that this ambiguous territory includes the elements that make possible the most complex properties of all, life and consciousness.

Crossing the line of metalloid ambiguity lands us fully and reasonably firmly in the heart of the Eastern Rectangle, the province of the nonmetals. Here lies some familiar territory, such as nitrogen, oxygen, and the halogens. Just to the south of nitrogen and oxygen lie two regions that have been known to nature, at least, for many years and exploited to the full. Immediately south of nitrogen is phosphorus, which was first isolated by the distillation and treatment of urine—an indication of the lengths to which chemists are prepared to go, or perhaps only a sign of the obsessive, scatological origins of their vocation. After this initial isolation, phosphorus was found to come in several varieties: white phosphorus, red phosphorus, and a black metallic kind. But this is not quite the lesson to note here. More significantly, it is to be noted how sharply properties change as we travel inland from the northern coast. The coastal element, nitrogen, is a colorless nonreactive gas, whereas immediately inland is phosphorus, a colorful reactive solid. A similar difference is found one step to the east, where sulfur, a yellow solid, lies just south of the colorless gas oxygen. There is an escarpment of difference between the coastal plain and the rank of land immediately to the south. In fact, all the regions of the northern coast from boron to fluorine are strikingly different from their immediate southern neighbors. Later we shall see that these two ranks of elements are fundamentally similar at a deeper level than this current superficial survey of appearance and utility, and that it is appropriate that they be placed in the location in the kingdom where they currently lie.

In this small area of the kingdom are the principal building blocks of life—carbon, nitrogen, oxygen, and phosphorus. Complexity can effloresce from subtly different consanguinity. We have already seen that phosphorus is a component of bone (calcium phosphate), but the region has been mined by nature for more than the scaffolding of vertebrates. Nature has discovered that phosphorus has a subtle personality that makes it particularly suitable for the deployment of energy in organisms. One salient characteristic of life is that it is not over in a flash, but involves the slow uncoiling and careful disposition of energy: a spot of energy here, another there, not a sudden deluge. Life is a controlled unwinding of energy. Phosphorus, in the form of adenosine triphosphate (ATP), turns out to be a perfect vector for the subtle deployment of energy, and it is common to all living cells. Viruses lack it, but acquire it from their hosts, and are sparked into activity once it becomes accessible. Here we see another organic alliance—between nitrogen, which is crucial to the utilization and conversion of energy, and phosphorus, which is crucial to its deployment, under the control of the proteins that nitrogen constructs. The central importance of phosphorus has prompted the agrochemical industry to set up major activities in this region of the kingdom, for the growth of crops depends critically on the ready availability of abundant phosphorus. Every single cell of the myriads cultivated each year requires it, just as we require it, and we obtain it from consuming what is grown.

The region of sulfur was also explored by nature—in nature's serendipitous, purposeless, but effective way—in an early investigation of the opportunities for life. Nature discovered that in some respects hydrogen sulfide (H_2S), the analog of water (H_2O), can be used by organisms in much the same way as water is used in the process of photosynthesis—as a source of hydrogen. The great difference to note is that when hydrogen is removed from a water molecule by a green plant, the excrement is gaseous oxygen, which then mingles with the globally distributed atmosphere. However, when hydrogen is removed from hydrogen sulfide in the interior of a bacterium, the excrement is sulfur. Sulfur, being a solid, does not waft away, so the colony of organisms has to develop a mode of survival based on a gradually accumulating mound of its own sewage. We still mine these ancient mounds of sulfur excrement from beneath the Gulf of Mexico. Sulfur's northern neighbor, oxygen, turned out to be a much more viable alternative to sulfur in nature's blind efforts to generate the transmittal and accumulation of information, and sulfur is now used only sparsely in organisms; and hydrogen sulfide is used only by primitive species that occupy a minor niche of nature.

That is not to say that certain modern organisms do not contrive to use sulfur for their own ends. Indeed, sulfuric acid, a compound of sulfur and oxygen, is the dominant product of chemical industry. There are very few manufactured products that have not been, at some stage of their fabrication, in contact with sulfuric acid. Sulfuric acid production has been used as an indicator of the output of a country's economy; moreover, this indicator is increasingly a sign of a country's agricultural vitality, since sulfuric acid is also heavily involved in the manufacture of fertilizers. There, its principal application is in the release of phosphates from the rocks in which they are embedded. Here, then, is another alliance, as carbon-based organisms use sulfur-based acids to manufacture phosphorus-based fertilizers to power nitrogen-based proteins.

South of this coastal area, the variation in properties is much less sharp, and the general tendency, as we have seen, is for nonmetallic character to give way to metallic character, as the land slips away to the southeast. Here lie regions that nature has hardly troubled to exploit: selenium, tellurium, polonium, and, slightly to the west, arsenic, antimony, and bismuth. Indeed, the region of arsenic, immediately south of life-giving phosphorus, is the source of a classic poison. Arsenic's activity as a poison stems from its close similarity to phosphorus, which enables it to insinuate itself into reactions that phosphorus undergoes but blocking their onward progress, while its subtle differences conspire to upset cellular metabolism. Arsenic is used to kill—benevolently, in certain drugs to combat infections, and malevolently, in nerve gases.

East of these regions lie the halogens: fluorine in the north to iodine in the south, and immediately below iodine the little-known astatine, on the southern shore. The halogens, apart from astatine, are enormously useful to nature and to industry, and these regions have been extensively exploited by both. Fluorine was largely a laboratory curiosity during the first incursions of chemical explorers into the region in the late nineteenth century; it is a particularly chemically virulent gas, and, indeed, is difficult to store, because of the ease with which it launches attacks on containers and transforms them into sieves. However, in the mid-twentieth century it became necessary to separate the isotopes of uranium for use in nuclear warfare and its relatively benign cousin, nuclear power, and the separation procedure made use of the volatile compound uranium hexafluoride. Fluorine was needed in abundance; thus the modern world found ways of handling the element, and it became available in abundance. One benign outcome was the use of fluorine as a hardening agent for tooth enamel and hence the overall improvement of the dental health of nations. The availability of fluorine also led to the production of fluorocarbons and among other things the functional convenience of the nonstick coatings they provide.

Chlorine, just south of fluorine, has been fully exploited by nature and industry, for it is found in great abundance in seawater. It is present there in the company of sodium as sodium chloride, the salt of the oceans, and—after a certain degree of refinement—the salt of the kitchen and table, too. Chlorine is also present in abundance in our bodies, whose fluids generally quite closely resemble the aquatic environment from which we emerged. Its role in our bodies, as in the ocean, is somewhat passive; it is present as a marriage partner of sodium, with little function of its own. Chlorine in its gaseous form, however, has strong similarities to its northern neighbor fluorine, being aggressively reactive, and chlorine gas has been used to kill both microbes and men. Chlorine may well be responsible for the killing of populations by more subtle means, for chlorocarbons and chlorofluorocarbons are used as refrigerants, and their consequent escape into the upper atmosphere has led to the depletion of the Earth's ozone layer. Ozone, a gaseous allotrope of oxygen, forms a protective shield around the planet which blocks the entry of harmful ultraviolet radiation from the sun—radiation that breaks delicate, functional organic molecules into useless and sometimes dangerous fragments.

Chlorine attacks ozone molecules and converts them back again into normal oxygen molecules, and thereby depletes the Earth's protective shield. There is much that is good in this region of the kingdom, but it has also been responsible for a great deal of environmental damage.

Further south is the lake—the liquid element—of bromine, the only lake of the Eastern Rectangle and one of only two lakes in the kingdom. Nature has largely bypassed bromine in its utilization of the elements, finding it adequate to employ chlorine, which is much more abundant, instead, and there has proved to be little that bromine's differences have needed to contribute. Human chemists, though, have found it a highly convenient element, for it can be attached to and plucked off organic molecules with ease, and hence used as a means to modify them to match industry's intentions. Bromine is essentially a utilitarian, backroom element, a region for the use of the manufacturer rather than the retailer. But even though rarely encountered, it is not to be scorned. One common application is in photography: silver bromide makes use of certain special photochemical properties of silver and bromine to help capture images of light.

Iodine is rather more useful to nature than bromine is. It differs significantly from chlorine, and nature uses it in certain biochemical processes as an internal mode of protection against invasion by foreign chemicals and organisms.

The narrow offshore Southern Island forms an unusual appendix of the kingdom. It is actually an extension of the Western Desert, and in certain maps of the Periodic Kingdom it is shown embedded in the mainland, forming a narrow component of the Isthmus of transition metals. The Southern Island consists of two largely uniform coastal strips. The northerly strip comprises a family of remarkably similar metals known as the rare earths or, more formally, the *lanthanides*. (Some still call them the "inner transition metals," a name inspired by the insertion of this island into the Isthmus.) The lanthanides are so similar to one another that until recently they could be separated only with great difficulty. Indeed, the near uniformity of their features suggests that it is not really worth making the considerable effort to separate them. Nature has seemingly no use for the lanthanides in its contriving of life, and humanity has only recently found certain sporadic uses for these regions. One of them is as a component of the phosphors that convert the energy of an accelerated electron beam into visible light of a variety of colors in a television tube.

The southerly strip of the island consists of elements termed the *actinides*. Until the effort associated with the development of the atomic bomb—the Manhattan Project, of the 1940s—the kingdom did not extend beyond uranium (except perhaps in distant stars). The Manhattan Project was effectively a land-reclamation project on the Southern Island; the discovery and manufacture of the so-called transuranium elements extended the kingdom and completed the southern strip of the island. A similar reclamation project is still in progress on the southern coast of the mainland, and every few years it is extended eastward by another region. There is little use for most of these latter-day regions, for they are so unstable that they survive only for very short times. In 1994, for instance, it was reported that one atom of element 110 had been manufactured—maybe.

On the mainland's eastern coast, there is another steep escarpment of chemical activity, from the halogens down to the noble gases at the shoreline. In due course we shall see why the great difference in properties is consistent with the fact that these two zones lie next to one another.

In geology there are reasons for the dramatic uplift of a terrain. In chemistry, too, there is a reason for the fierce uplift in activity from the shore of the noble gases to the heights of the vigorously reactive halogens. We must for now enter it into our sack of problems to resolve.

That the noble gases are largely aloof from all chemical activity does not mean that they are unproductive zones of the kingdom. In some cases, it is precisely that lack of activity which renders them useful, since they can be used to provide an inert local atmosphere when the ordinary atmosphere—or even an atmosphere only of nitrogen—would be too reactive. The noble gases have certain other physical properties that render them useful. One is the remarkably low boiling point of helium, which makes it a useful refrigerant when exceptionally low temperatures are sought. Another is the colorful display obtained when an electric discharge is passed through the gases—a phenomenon observed in the generically named neon light.

There are few regions of the kingdom that have no application and have not been exploited by nature or industry. Here we have pointed out only the principal uses and occurrences of the elements: a comprehensive list would turn this traveler's guide into an encyclopedia of biochemistry, mineralogy, and industrial chemistry. But just as there are regions of the real world that have not contributed much to the global economy, so there are regions of the kingdom that for one reason or another are quiescent. Every element has at least a minor variation in personality that renders it useful in principle, and perhaps more so than its neighbors in the kingdom, for a particular task. So there must be reasons for the lack of exploitation other than the intrinsic characteristics of the element itself.

There are two, typically. One is that the element exists on Earth only in very small quantities, and is almost more of an idea than it is a stockpile of goods. Such is the case with francium and astatine, each of which occurs in such submicroscopic quantities that there is no possibility of commercial exploitation. It has been estimated that in the whole of Earth at any one time there are only about seventeen atoms of francium. The transuranium elements in the recently reclaimed areas of the southern shore are prepared in similar quantities, and on account of their extreme rarity are unlikely ever to be exploited. The second reason is radioactivity. Bismuth, in the deep southeast, is the kingdom's outpost of stability; all the elements beyond it are radioactive. The southern strip of the offshore island (the actinides) and the southern fringe of the kingdom's mainland are dangerous places, for wherever one is there is radioactivity. Although radon is abundant, for instance, there is little incentive to develop the properties of this dangerously radioactive gas. All along the southern shores, there are skull-and-crossbones signs, and here even the interest of chemists begins to wane and wariness quenches curiosity.

CHAPTER 3

PHYSICAL GEOGRAPHY

Many features of the kingdom are not apparent to casual visual inspection, just as is the case with the terrain of actual countries. To achieve deeper knowledge of the regions and of the underlying rhythms of the kingdom it is necessary to make measurements. Some of these measurements are straightforward; others require greater sophistication. However, in all cases we can attach numbers to each region and imagine the variation in number as an aspect of the terrain. Exactly the same is done in mapping real landscapes, where altitude is measured and then displayed as color or modeled as the actual rise and fall of the land. Color or contour may also be used to indicate the variation of other properties, such as the density of population or the acidity of the soil.

In this chapter, we shall explore the ghostly secondary images of the landscape, where the changes in various properties are depicted as changes in altitude. As this is in any case an imaginary kingdom, a land of chemists' dreams, there need be no constraint on its portrayal except truth, and to the mind's eye the terrain may rise and fall according to the features we seek to portray. We have already seen an example of this dreamlike modulation of the landscape in relation to the rise and fall of chemical activity, where a steep eastern escarpment plunges down from the plains inhabited by the halogens to the littoral of the noble gases. Earlier we also invoked altitude when we depicted the northern shore as quite different in physical and chemical character from the southerly uplands. There was nothing quantitative in the depiction; it was simply evocative symbolism. Now, though, we intend to depict actual numbers obtained by well-defined physical measurements. The altitude and depth of the imaginary landscape is now much more real, though not an actual altitude or depth.

Full exploration of the physical geography of the kingdom requires us to take another important step. So far, we have surveyed the kingdom largely from the skies and have seen the regions and their undulating rhythms from afar. We have assayed the phenomenology of the elements, their appearance, form, color, and physical state. Now we must land. Wherever we land, we can examine the detailed texture of the terrain, and through the imaginative eye readily replicate what actually required detailed observation and precise measurement. That is, we can visualize the atoms from which the elements are made. It is the identity, form, and structure of the atoms of a region that distinguish one region from another.

The atoms will be the topic of much more detailed discussion later in this account of the kingdom, for they and their insides are the basis of complete explanation. For the time being, all we need do is picture them as a shinglelike texture, with every pebble of a given region exactly the same as all the others of that region but different from the pebbles of any other region. In this kingdom, microscopic texture is the foundation of function and the distinction of feature. In particular, atoms are distinguished by their masses and their diameters, as well as certain other characteristics that close inspection will unfold.

First, we consider a straightforward property of atoms—their masses, which lie between about 1 x 10^{-31} kilograms and 1 x 10^{-29} kilograms. It is far more convenient to discuss masses in terms of their relative values, and for convenience we shall set the mass of the atom of hydrogen as 1 and refer all other masses to that value. Thus a carbon atom, which is about 12 times heavier than a hydrogen atom, is assigned a mass of 12, and a uranium atom, which is about 238 times heavier, is assigned a mass of 238. Modern chemists have refined this relative scale, but their numbers, which are often termed *atomic weights*, turn out to be very close to those we obtain in this more primitive way.

Now, imagine with your mind's eye the terrain of the kingdom in terms of altitudes that represent the relative masses of the atoms of the elements (fig. 3). To picture it, you can think of the landscape as sloping up from the northwest tip of the kingdom to the dangerously radioactive regions of the far southeast. The Southern Island slopes up from west to east, too, and the southernmost strip of the island is uniformly higher than its northern strip. Someone traveling due east anywhere on the mainland will find the land rising. Someone traveling due south in any region of the kingdom will find the landscape rising sharply. The lowest altitudes are in the northern outpost of hydrogen, at the Northeastern Cape of helium, and at lithium on the Northwestern Cape. The altitude rises uniformly almost everywhere, and a traveler on the southern coastline will be over two hundred times higher than a traveler on the low-lying, sea-lapped northern coast. In this peculiar vision of the kingdom, the reclaimed new lands of the far southeast spring high into

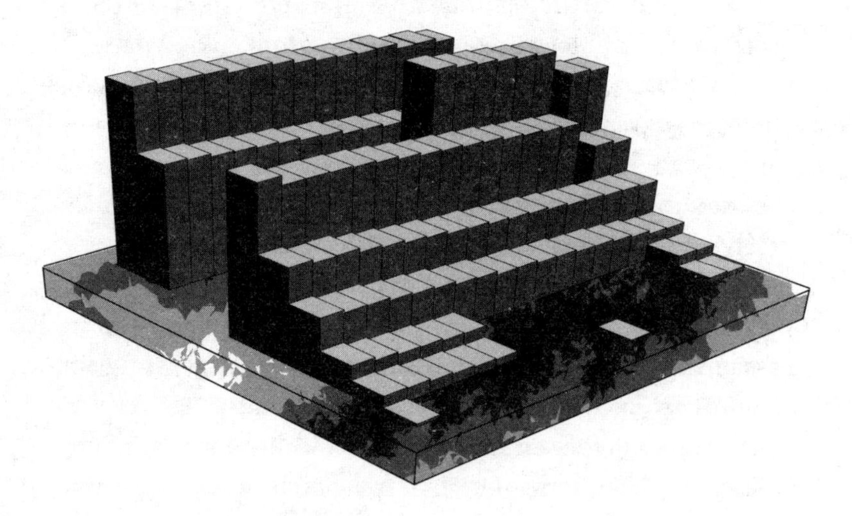

FIG. 3.

The kingdom plotted in terms of the masses of the atoms. We are viewing the kingdom from the northeast corner. The Southern Island lies in the distance. Hydrogen, in the right foreground, is barely visible above the waves.

....

the air as they are discovered, forming the headily high, precipitous cliffs of the radioactive southern coastline—higher even than the heights of the Southern Island.

There are a few places where this trend in altitude will trip a careless walker. Here and there—in the Eastern Rectangle, between tellurium and iodine, and in the Isthmus, between copper and nickel—a dreaming foot will trip, because here the land falls away very slightly instead of rising. These tiny faults in the landscape clearly need the equivalent of a geologist's explanation, and for now we shall let them stand in the line of observations requiring interpretation. More to the point, the fact that there do seem to be tiny faults in the landscape is a signal to the alert that the masses of the atoms cannot be a fundamental characteristic of the elements. A truly fundamental property of atoms—one to which we should be able to correlate all other properties both physical and chemical and so absolutely interpret the kingdom—would not be a property that admitted any faults at all. Clearly, the masses of atoms seem closely correlated with whatever is truly fundamental, because in general the landscape rises so smoothly, but they themselves cannot be fundamental. Nevertheless, for pioneers in the kingdom, it is certainly useful to know that the atoms of a southeastern region are heavier than the atoms of a region to the north and west. They know that to go south or east they need to climb; and they know that the track back to the Northwestern Cape is downhill all the way.

The atoms of a region are also characterized by their diameters, although the variation in the diameters of atoms is far less than that of their masses; an atom of the heavy element uranium is only two or three times the diameter of an atom of the lightest element, hydrogen. The diameter of a typical atom is about 0.3 nanometers, and a nanometer (1 nm) is one-billionth of a meter, or one thousand-thousandth of a millimeter, which is just on the edge of our ability to imagine. So, the landscape draped in altitude denoting diameter is much flatter than the landscape of atomic masses (fig. 4). It is also much less regular in form, which immediately tells us that a fundamental property of atoms is not to be found in their diameters. Nevertheless, with the variation in diameter mapped onto the landscape, much becomes understood, as we shall see.

FIG. 4.

The kingdom plotted in terms of the diameters of atoms. This view is from the same position as in fig. 3. The diameters are based on the lengths of the bonds the elements form, so the noble gases are not ascribed values in this view because they do not form bonds.

....

Broadly speaking, the landscape of diameters rises from north to south and falls from west to east, but there are many exceptions. It is rather contrary to intuition, at first sight, that as atoms get heavier from northwest to southeast they also get smaller. A particularly important instance of this apparently bizarre trend is from west to east across the Isthmus, where the land sinks as we travel east and then climbs again into the Eastern Rectangle, where the fall resumes. Something is clearly at work in the inner geology of the kingdom for the land to droop so significantly, and this is another candidate for explanation.

There is an additional peculiarity in this landscape. Although the landscape of masses rises almost without exception as you go from north to south, the rise is much less noticeable in its southern parts. Indeed, inland from the southern coast, there is even a slight regression, and the regions close to platinum and iridium are lower than the lands immediately to their north, a trend contrary to the trend of their masses. Clearly, the landscape of diameter is portraying complex subterranean influences that we cannot even speculate about at this stage but that have profound influences on the nature of the terrain and can therefore be expected to affect the properties of the elements. Nevertheless, this complex variation should not conceal the fact that the variation in height is not random. The landscape is one of smooth trends—rising valleys, drooping and dimpled plateaus, not randomly juxtaposed gorge, gully, and peak. There is an underlying geology at work; the kingdom has here, as elsewhere, a rhythm, and the juxtaposition of the elements has an underlying justification.

We now make contact with everyday reality by springing back from this consideration of atoms into the real world and considering the density (the mass per unit volume) of the elements (fig. 5). We shall look in particular at the densities of the metallic elements of the Western Desert. We imagine a sheet laid over the Western Rectangle and along the Isthmus, with its height above the surface representing density. Generally speaking, the altitude of this imaginary sheet sweeps upward from the Northwestern Cape (at lithium) toward lead on the southeast fringe of the desert. The rise is not uniform, and a particular feature is the high mountain ridge which peaks at iridium and osmium. These two are the densest elements of

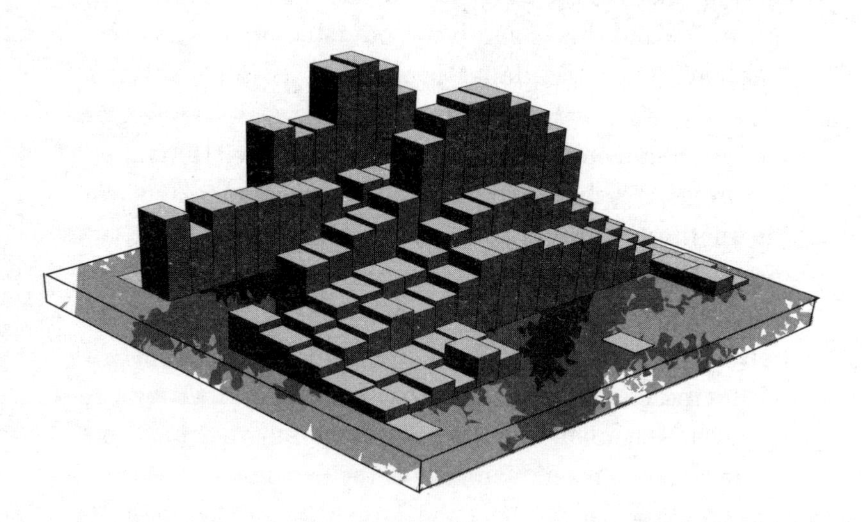

FIG. 5.

The kingdom plotted in terms of the densities of the elements as solids (including solidified gases). Note the very high densities toward the south in the vicinity of iridium and osmium.

....

all, each with a density close to 22 grams per cubic centimeter; even lead, the popular paradigm of dense metals, has a density of only 18 grams per cubic centimeter. In contrast, the density of magnesium, far to the north, is only 3 grams per cubic centimeter.

We have now assembled enough knowledge of the physical geography of the kingdom to embark on the first stages of an interpretation of this variation of a tangible property, the density. We have seen that masses increase from northwest to southeast across the kingdom, and in particular across the eastern part of the Western Desert. However, we have also seen—but have not explained—

that although atomic diameters do vary, they do not vary by much, and in particular they do not markedly increase from north to south, as masses do. Great mass packed into small volume corresponds to high density, so we should expect the elements of the southern coast and immediately inward of it to be particularly dense, just as observed. Of course, deep understanding of this variation in density will come only after we have understood the reason for the variation in diameter. Nevertheless, we have taken a step characteristic of science—the identification of the point to explain in order to account for a phenomenon.

We have concentrated on the Western Desert for this first little foray into physiographical explanation because its regions are solids, made up of spherical, pebblelike atoms that pack together compactly and in nearly the same patterns. Although the atoms of a given element might arrange themselves in a slightly different way from the atoms of another element—some with, say, eight nearest neighbors, some with twelve—nevertheless, the variation in packing patterns is quite small. The elements of the Eastern Rectangle, in contrast, pack together in a much wider variety of ways, and typically have a smaller number of immediate neighbors and much more open, frameworklike structures. It is consequently much harder (though still possible) to correlate the density of the elements of the Eastern Rectangle with the diameters and masses of their atoms.

Our physiographic eye will now don other spectacles and view the kingdom in a different light. Here we turn from the masses and dimensions of the pebbles of the terrain to a certain variety of change that the pebbles can undergo. Chemistry, after all, is the science of changes in matter, and so it is inevitable that we examine this aspect of the kingdom. At this stage of our travels, though, we shall confine our attention to the most primitive change an atom undergoes, and map this characteristic over the kingdom. Once again, we shall see a rhythm, and in due course this rhythm will show us how even complicated types of chemical change are reflected in the layout of the kingdom. In particular, we shall begin to establish a view of the kingdom that lets us quantify carbon's potent mediocrity and see what renders it essential to life.

Each of our forays into the kingdom at ground level equips us with a new concept. On first landing in the kingdom, we found that the texture of the ground introduced the concept of the atomic structure of the elements. Now, to penetrate the substructure of the terrain, we need the concept of *ion*. An ion is an atom that has gained or lost electrons and has thereby become electrically charged. Electrons are negatively charged fundamental particles that are largely responsible for an atom's chemical properties. When an atom loses electrons, it becomes positively charged—with one or two or three units of positive charge depending on whether it has lost one or two or three of its electrons. Positively charged ions are known as *cations*. When an atom gains electrons, it becomes negatively charged. One additional electron endows it with one unit of negative charge, two additional electrons means two units of negative charge, and so on. Negatively charged atoms are known as *anions*. These names were coined by Michael Faraday in the nineteenth century when he was establishing the effects of electric currents on solutions of ions. He found that there were two types of ions that traveled in opposite directions in the presence of electrodes—one toward the positive electrode and the other toward the negative electrode. *Ion* is the Greek word for "traveling"; "cat" is derived from the Greek for "down" and "an" from the Greek for "up".

The landscape we shall now construct in our imagination will show the ease with which atoms form ions. As so much of chemistry involves the formation or the incipient formation of ions, this landscape is very close to the depiction of actual chemical properties. As we shall see, there are complex and subtle rhythms of the land, yet they are explicable rhythms, not random variations.

The energy required to form a positively charged ion—a cation—from an atom of an element is called the *ionization energy* of the element. Ionization energies can be reported in a variety of units, but the most convenient for our purposes—because it leads to a ready visualization of magnitudes and gives numbers not far removed from 1—is the electronvolt (eV). One electronvolt is the energy needed to drag an electron through a potential difference of 1 volt (1 V). When an electron moves from one electrode to another in a typical 1.5-volt cell, such as that used in a cassette recorder or small portable radio, it releases an energy of 1.5 eV, and in a 12-volt automobile battery, it releases an energy of 12 eV. The ionization energy of an atom of hydrogen is 13.6 eV. To visualize this energy we can think of the interior of the atom as having an electrical potential of 13.6 V, and points far removed from the atom, to where the ejected electron goes, as having a potential of 0 volts. This ionization energy is about middling in value; typical values range from 4 eV (a four-volt difference) to 15 eV (a fifteen-volt difference). The energy required to remove a second electron is always higher than that required to remove the first electron from an electrically neutral atom, and the energy required to remove a third electron is even greater. To keep matters simple, we shall consider only the *first ionization energy*, the energy required to remove the first electron from a neutral atom.

The altitude of the landscape of ionization energies ranges in height from about 4 eV (at cesium) to 25 eV (at helium). There is one broad trend that is easily discerned, and a variety of ripples, dimples, and excursions that need to be noted (fig. 6). The broad trend is that ionization energies generally increase from west to east across the kingdom and decrease from north to south. For instance, on the western seaboard of the alkali metals, the ionization energy decreases from 5.4 eV at lithium to 3.9 eV at cesium. Here is an apparent chemical correlation, for we have seen that when it rains on this coastline the reaction varies from gentle seething at lithium to explosive violence as we go south to cesium: the vigor of the reaction is closely related to the ease with which the atoms can let go of an electron. This trend can also be related to the trend in diameters of the atoms, for as the atoms become larger from lithium to

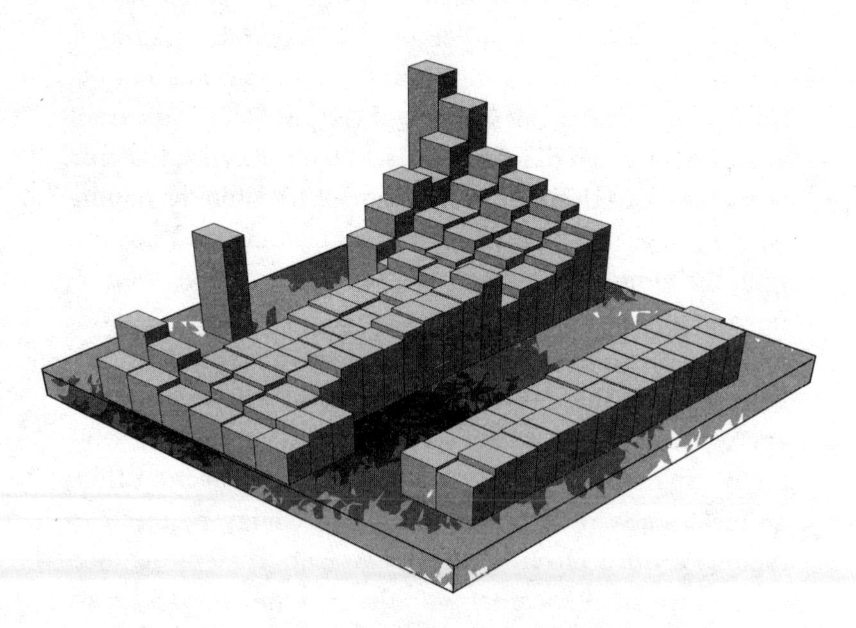

FIG. 6.

The kingdom plotted in terms of the ionization energies of the elements. This view is from the southwest. The three highest peaks in the distance are fluorine, neon, and helium.

cesium, so it becomes easier to extract an electron. The rhythms of the kingdom are beginning to form a harmony in place of a series of uncorrelated melodies.

The trend of increasing ionization energy from west to east is not entirely uniform, but even to the casual eye the regions of the Eastern Rectangle are all significantly higher than those of the Western Rectangle. The most complicated variations occur across the Isthmus, and the complexity of the chemical reactions of these elements reflects this variation, as we shall see. But even across the Isthmus the general rise is in line with the transitional character of these metals.

The low terrain of the Western Desert should not come as a surprise. Everyone knows that a feature of metals is the readiness with which they conduct an electric current, which is a stream of electrons passing through a solid. For this to be possible, at least some of the electrons of the atoms need to be mobile—that is, the atoms that lie packed together in the solid release electrons to a common pool, that lies like an ocean among the resulting cations. We should think of the pebbles in this metallic terrain not as atoms but as cations, bathed in and held together by a pervading sea of electrons. These electrons can respond to an applied electric field. If that field is applied by a source of constant potential difference (colloquially, voltage), as is the case if the metal is connected to the electrodes of a battery, then the electrons flow and form a current. If the electric field is that of an incident ray of light—that is, electromagnetic radiation, an oscillating wave of electric and magnetic fields—the loose electrons respond by oscillating in harmony with the radiation, and in so doing generate a ray themselves. Thus when light hits a metal, we see a lustrous surface. When that surface is flat, the incident pattern of radiation is reproduced in the generated radiation, and we see a mirror image. Our own image on a smooth metallic surface is a portrayal of the ripples in the mobile electron sea caused by the rays that have reflected from us.

To be metallic requires, in a sense, a looseness of electrons. As the ionization energies rise on traveling east, the Western Desert gives way to the nonmetals of the Eastern Rectangle, where the ionization energies are too large for electron loss to be feasible. A part of the reason for the rising ionization energies is the decreasing diameter of the atoms as we travel east; for some reason, it becomes increasingly difficult to remove electrons from the compact atoms found to the east in the kingdom. Now we have another correlation with size, and, once again, if we could understand the variation in the diameters of atoms then we would be well on the way to understanding the variation in ionization energy.

While we cannot readily expect to find cations far to the east, because the ionization energies of the atoms are too high and electron loss is consequently too difficult, perhaps we can expect to find anions (negatively charged, electron-rich ions) here instead. Anion formation is a rather more subtle process than cation formation, but a helpful measure of its feasibility is the energy change that occurs when an electron attaches to an atom. This energy change is called the *electron affinity* of the element, and it is also conveniently measured in electronvolts. An atom that has a positive electron affinity releases energy when it acquires an additional electron; an atom with negative electron affinity must be supplied with energy to overcome its objection to acquiring an additional electron. Some elements, such as the halogens, have positive electron affinities; some, such as magnesium and helium, have negative electron affinities. *All* elements have negative electron affinities when it comes to the attachment of a second or third electron; it requires energy to attach an electron to an anion, because an anion and an electron are both negatively charged, and like charges repel. So, just as we did for ionization energies, we shall consider only an element's first electron affinity—that is, the energy change that accompanies the attachment of an electron to a neutral atom.

The landscape depicting the variation of electron affinity is much less regular than the other landscapes we have seen. In places where electron affinities are negative, it plunges below sea level; in places where affinities are high—that is, where energy is released when anions are formed—it rises sharply. Nevertheless, here too a trend can be discerned: although the chasms and peaks appear to be randomly distributed, the chasms occur toward the western shore and the peaks occur toward the northeast, close to fluorine. The noble gases of the eastern littoral, with their mountainously high ionization energies, have negative electron affinities, at least in the northern regions, so the mountains marking the halogens plunge sharply downward. However, the principal feature to note in this landscape is the cluster of mountains in the northeast, and particularly the far corner made up of nitrogen, oxygen, fluorine, and chlorine. These four are the elements most avid for electrons, where it is most likely that anions will form.

Here, too, is a loose correlation with atomic diameter: high electron affinity correlates—generally, at least—with small, compact atoms, which are found in the northeast of the kingdom. To understand anion formation—and wider aspects of the attachment of electrons to atoms—we need to understand why there is this loose correlation with size, and what the variation itself means.

This survey of the texture of the land and the physiography of the regions has shown us that there are correlations between the location of regions and certain of their properties—that the kingdom is not a patchwork of casually juxtaposed autonomous regions but a manifestation of underlying currents that effloresce at the surface as properties. It may help to return to higher altitude now and summarize these different views of the kingdom. The landscape depicting atomic mass correlates most closely with location, increasing from northwest to southeast, with only an occasional small exception. Next, with respect to atomic diameter, the elements generally increase toward the south and become smaller from west to east. Although atomic diameter varies much less than mass, it does so in a more complicated way: superimposed on these smooth large-scale trends are local falls, dimples, and bulges. A further important point is that the elements of the southern coast and their inland neighbors are smaller in diameter than simple extrapolation from the north would suggest. The densities of the elements of the Western Desert generally increase from northwest to southeast, and the greatest densities are found in regions where atoms are anomalously small—that is, the southern coastal elements and their inland neighbors. And finally, properties related to energy mirror, more or less loosely, the variation in the diameters of the atoms. Ionization energy is low for metals, and lowest of all in the far southwest, close to cesium. They are highest for elements close to the northeast corner, and we cannot expect these elements to form cations readily. Electron affinities vary in a much more complicated way, with high values often next to low or even negative values. Nevertheless, there is a general tendency for the highest values to be found in the far northeast, so that although these elements do not form cations, we can expect them to form anions.

PART 2

HISTORY

CHAPTER 4

THE HISTORY OF DISCOVERY

Many chemists, physicists, and artisans have contributed to the discovery of the regions of the kingdom. Some have stumbled on a new element unexpectedly; others have planned their journeys of exploration, in the expectation of discovering an element whose existence they already suspect. Some of this exploration is the equivalent of land reclamation, for it is along the southern shore of the mainland that there is every expectation of creating new land.

The discovery of many elements is lost in antiquity. Which genius first isolated copper and opened the way to the progress of civilizations? Who first identified iron, the element that consolidated that progress and cut more forcefully forward? No one knows. The names of the Drakes, the Magellans, the Cabots, and the Cooks of chemistry come down to us only from the seventeenth century onward; earlier explorers, like the discoverers of ancient lands, remain anonymous. Even the names of some regions have origins that are effectively unknown or open only to speculation. More recently, the naming of the regions has been the privilege of their discoverers (although this privilege is being usurped by committees), and the etymology of the name of a new element has been better documented. There are still a few squabbles, for it is sometimes not entirely clear who first rightfully laid claim to a particular part of the coastline of the kingdom, and committees are attempting to resolve these local conflicts.

Broadly speaking, the discovery of new elements has depended on the development of new technologies, and these newly discovered elements have in turn given rise to even newer technologies, as iron was necessary for the discovery of titanium. The first enabling technology was fire, which drove compounds apart in ways that were at first incomprehensible and verging on the apparently magical. Thus iron became available in bulk only when fire was applied to certain stones. Fire itself, when carefully controlled and no more vigorous than resulting in a gentle charring, led to the isolation of another element, carbon (as charcoal), which in turn artisans found could be used in combination with fiercer fires to react with other stones, thereby freeing new metals. Tin, lead, and other regions of the Isthmus in the neighborhood of iron were discovered in this way, and in turn made new discoveries possible.

Most discoveries in the kingdom have not been of the lazy, serendipitous kind, where an explorer merely stumbled into a new region, like a prospector panning for gold or a caveman picking up a lustrous stone or a steely meteorite. Rather they have been the prize of laborious investigation, drawing on the power of new technology to press forward the frontier of human knowledge. These technologies have been used to isolate new elements that were formerly found only in combination. It was once difficult to recognize a new substance as an element, and there have been mistakes. Modern techniques have removed the ambiguity of this process. Now we can take a sample, atomize it, weigh the atoms, and determine whether or not they are all essentially identical—that is, whether or not the sample is that of a single element. Long ago, inference from qualitative data was the principal tool, and it is a remarkable monument to the power of human reason that so much of the kingdom could be discovered in this way.

A few elements—most evidently gold, copper, and sulfur—can be found in native form, and among these elements are the gases of the Earth's atmosphere. However, it required some sophistication to realize that this invisible envelope was a mixture of elements (and of some of their compounds, too), and this understanding came quite late in the history of the kingdom. The simplest of elements, hydrogen, is found hardly at all in its native state on Earth, except for pockets of pure hydrogen gas trapped below ground by rock formations. Hydrogen is the most abundant element in the universe, however, and all other elements—apart from the abundant but inert helium—can be regarded as a minor (but crucial) contamination of this universal substance. The cosmic ubiquity of hydrogen was not recognized until early in the twentieth century, when, contrary to philosophical prediction, astronomers developed tools—various forms of spectroscopy, in which the characteristics of radiation absorbed or emitted by substances are monitored—that enabled them to determine the composition of extraterrestrial bodies.*

Oxygen was discovered in 1774, by making use of a distant nuclear fire, the sun. Joseph Priestley, the eighteenth-

*The positivist philosopher Auguste Comte asserted in 1835, in relation to the sun and other celestial bodies, that "we understand the possibility of determining their shapes, their distances, their sizes and motions, whereas never by any means will we be able to study their chemical composition, mineralogical structure, and not at all the nature of the organic beings living on their surface." Such is the reliability of philosophical speculation from armchairs.

century English chemist and nonconformist clergyman, used a lens to focus the heat of the sun onto a phial of mercury oxide and generated bubbles of a life-giving gas. Priestley is perhaps best regarded as the *last* person to discover oxygen, for others had reported its preparation but had not realized that it was an element. The Swedish chemist Karl Scheele in fact discovered oxygen two years before Priestley, but delays in publication lost him the priority he deserved. As in the discovery of terrestrial lands, it is typically the last person to "discover" a region who is remembered by posterity. Hydrogen had been prepared long before it was recognized as an element by the eccentric misogynistic recluse and chemist Henry Cavendish (from whose fortune the Cavendish Laboratory at Cambridge was later to spring). With hydrogen and oxygen identified, it was an easy step to argue (as Cavendish did in 1781) that water, which is formed when the two gases are sparked, is not itself an element.

At the beginning of the nineteenth century came a very special technological advance, which is still in widespread commercial use today. *Electrolysis* is the breaking down of matter by the passage of an electric current through it. After the development of voltaic piles and other devices that generated steady currents from chemical reactions, it was only natural to see what could be achieved by the then novel phenomenon of electricity. It was known that electricity could shock; what could be more natural than to find out whether electricity could shock matter into different forms? In 1807, at the newly founded Royal Institution of Great Britain in London (the world's first custom-built laboratory), Humphry Davy applied an electric current to molten caustic potash, which we now know to be potassium hydroxide, and obtained beads of a silvery reactive metal that he named potassium. This metal, as we have seen, is located on the western shore of the Periodic Kingdom; within a few days, Davy had the wit to apply the same technique to a similar compound, caustic soda (sodium hydroxide), and obtained the northern neighbor of potassium, which he named sodium. Modesty prevailed in those distant times: were modern modes a guide, he would probably have named them royalinstitutium and londonium, or even humphryium and davyium.

Soon electrolysis was applied to all manner of substances. Magnesium was culled from its compounds by Davy in 1808, calcium in 1808 (also by Davy), and strontium in 1808 (Davy again). The greatest recorded surge in knowledge of the kingdom and extension of its regions since history began flourished under the impact of this important technique, and the kingdom grew under a multitude of hands throughout Europe. Explorers mapped the eastern promontory of the halogens, bringing back knowledge of chlorine (discovered by Scheele in 1774 but named and recognized as an element by—who else?—Davy in 1810) and bromine (discovered not by Davy but by Antoine-Jérôme Balard in 1826). By the 1860s, about sixty of the kingdom's regions were known. Fluorine, the peak of the mountain range of reactivity, remained uncharted by explorers until late in the nineteenth century, but this Everest, too, eventually succumbed to electrolysis and was added to the kingdom by Henri Moissan in 1886.

One advantage in having a chart like the kingdom is that it suggests where other regions ought to lie, and hence where explorers should plan their treks. Moreover, because the kingdom is not a random jumble of regions but a land of consanguineous neighbors, the characteristics of a missing region can be predicted—in broad outline, at least, and in certain aspects of detail—by noting the characteristics of its neighbors. The Russian chemist Dmitri Mendeleev, who is one of the principal historical figures of the kingdom and who in 1869 essentially laid it out in its current form, showed the power of this approach to discovery. For example, he recognized a gap close to silicon, which, for want of a name, he called *eka*-silicon (*eka* being the Sanskrit name for the numeral 1), and predicted its properties on the basis of his knowledge of what that location required and what he knew of silicon. When this region (germanium) was discovered by the German chemist Clemens Winkler in 1886, it was found that Mendeleev had been substantially correct. The same was true for *eka*-boron (scandium, not isolated until 1936, although it had been known since 1876) and *eka*-aluminum (gallium, isolated in 1875). Marie Curie, that intrepid female explorer of the dangerous southern shore of the kingdom, in collaboration with her husband, Pierre, was led to her isolation of radium from tons of pitchblende, a uranium ore, by her anticipation of its likely properties, based on her knowledge of barium, its northern neighbor.

This process of progressive exploration is all very well, but it can sometimes fail. And fail it did in one spectacular instance in the Eastern Rectangle. As we have seen, at the kingdom's western shoreline there is a precipitous cliff, formed by the highly reactive alkali metals, which juts upward from the sea, and inland of this cliff is the slightly lower mountain range of the alkaline earth metals. These western peaks of reactivity were mirrored on the eastern fringe of the Eastern Rectangle—by the reactive regions of oxygen and sulfur and, just beyond them, the high mountain range of the halogens, which, it was thought, plunged down to the empty sea. But this symmetry of the kingdom was an illusion, because at the base of the halogen cliffs lay the littoral of the rare, later inert, and now noble gases.

Here, then, was a surprise! In 1894, the chemist William Ramsay and the physicist Lord Rayleigh noted that there was a difference in the density of nitrogen gas obtained by the decomposition of nitrogen compounds and the density of the same gas obtained by isolating it from the atmosphere. (It should be said that the phenomenon was not entirely unheralded; as long ago as 1785, Henry Cavendish had speculated that there might be a nonreactive gas mixed with atmospheric nitrogen, but his idea had not been pursued.) Rayleigh thought the difference in densities could be accounted for if the nitrogen from the decomposed compounds included an unknown lighter substance; Ramsay took the opposite view, conjecturing that the atmospheric nitrogen was contaminated by a heavier gas. In due course, he found that he could separate the atmospheric "nitrogen" into nitrogen and another gas, which was far less reactive; thus he discovered argon (the name comes from the Greek word for "lazy"), the first region of the low coastal fringe to make its appearance. As we have remarked, argon is far from rare, being even more abundant in the atmosphere than carbon dioxide is; argon was just unexpected, the terrain of the kingdom giving no hint of its existence. When we turn to the kingdom's substructure—the fundamental principles of its organization—we shall see that argon could have been predicted: it is essential for the subterranean pattern. But in 1894 that pattern was unknown, and the shoreline of the noble gases was a Lost World. Incidentally, the subterranean pattern forbids the existence of a matching low-lying Lost World beyond the western precipice, so young explorers are advised not to waste their time, and older ones their speculations, on what might lie there. There is quite probably an Atlantis offshore, but it certainly does not lie to the west; read on.

The discovery of argon led to the realization that there were more regions in the area, and neon, krypton, and xenon were soon discovered, all of them by William Ramsay in 1898. Helium, the lone promontory in this strand of regions and its northernmost point, had already been discovered—not on Earth but in the sun, hence its name. The discovery of helium is an extraordinary story. In the first place, helium makes up about 25 percent of the universe, yet it lurked invisible and unnoticed until 1868, when it was observed spectroscopically during a solar eclipse. Second, despite this abundance helium was not isolated in the laboratory until 1895, when Ramsay, in pursuit of a report that an unfamiliar gas had been discovered by heating uranium ores, isolated the gas. Finally, although it was first identified in the hottest part of the local cosmos, helium is in fact used in the technology of the coldest parts. Liquid helium, which forms when helium gas is cooled and squashed into submission, is essential to cryogenics and cryotechnology, and until recently it was the only route to the achievement of superconductivity.

By the turn of the century, just one region of the eastern shore remained to be discovered. Once again, it was William Ramsay who beat the path to the region, isolating the radioactive gas radon in 1908. Of course, Ramsay had his collaborators; there were independent and disputed discoveries, and Ramsay himself worked with a team of other chemists, but no other complete strip of land of the kingdom owes so much to a single person.

Another surge in the mapping of the kingdom occurred in the mid-twentieth century, under the pressure of war. The Manhattan Project of the 1940s, the aim of which was to construct an atomic bomb, turned out to be a giant reclamation project as well, extending the Periodic Kingdom outward along its southern shore as new elements were made and recognized. First, it extended the southern strip of the offshore island in the southern sea, as the metals collectively known as actinides were discovered. At the start of the Manhattan Project, the Southern Island, as has been noted, terminated at uranium. That element, though, had the potential to beget new elements when appropriately organized in what were then evocatively called piles and are now called nuclear reactors. Uranium begat neptunium and plutonium, and the island's south coast began to grow toward the east. Investigation of the northern strip of the Southern Island—the lanthanides—benefited from this extension of the southern strip. Special techniques had to be developed to separate the actinides, which were almost identical, and the application of these techniques—particularly *chromatography*, in which species are separated according to the time it takes them to travel through a sticky medium—to the northern strip greatly facilitated the separation of the lanthanides, too.

More recently still, reclamation has proceeded by synthesis—that is, the fabrication of elements from simpler components. The highly transient elements of the mainland's southern shore—dubnium, joliotium, rutherfordium, bohrium, hahnium, and meitnerium—have been made in machines that are another outcome of the Manhattan Project: cyclotrons, synchrotrons, and linear accelerators. To make a few fleeting atoms of one of these essentially useless elements is to claim a whole new shoreline for the kingdom, and perhaps to form varieties of matter that exist nowhere else in the solar system. To make new atoms, atom is hurled against atom. If the collision is successful, they blend and for a fleeting fraction of a second cling together to form an atom of a new element. It is near these southern outposts of the kingdom that an undiscovered Atlantis might lie. Current scientific opinion holds that although the southern shore consists of fleeting regions, further out in the ocean of instability there may lie an Island of Stability, where as yet wholly unknown elements exist and perhaps survive for longer than those we now struggle to prepare. Almost certainly that island, when reclaimed, will be not only uninhabitable because of radioactivity but largely useless because of the brevity of its regions' lives, for the atoms of these new elements might last only a few months. But the chance that they may be there is enough to inspire the nuclear Columbuses to set sail toward them. The knowledge alone would be priceless.

CHAPTER 5

THE NAMING OF THE REGIONS

Once exploration of the kingdom became a science, the nomenclature of the regions became systematized to a certain extent. As we have remarked, the naming of a region, with certain constraints of propriety, has been the privilege of its discoverer. The constraints are those largely of common sense. Triviality (mickeymousium), blasphemy (godium), and obscenity (****ium) are avoided and would not be accepted by the scientific community anyway, but certain witty and perhaps unnoticed and unintended jokes have slipped in, as with gallium, named by its discoverer, François Lecoq de Boisbaudran, supposedly for France (from the Latin *Gallia*) but probably more for himself (from *Gallus gallus*, the cock). These days, committees tread more forcefully in the corridors of the kingdom and aspire to regulate the process of nomenclature. This aspiration is particularly acute now that so much of a nation's scientific budget can be expended on land reclamation off the southern coast, the only place where there is room for new names (though an occasional tinkering with established names still takes place elsewhere in the kingdom).

An inversion of history might be helpful here. In recent years, a scheme of nomenclature has been devised that depersonifies the regions and makes their naming a systematic procedure rather than a vehicle of pride and a personal passport to a variety of immortality. Here, as beneath the tables of so many committees, poetry and pride lie trampled, discarded for a more prosaic mode: unium for the first element (hydrogen), biium for the second (helium), up to decium for the tenth (neon), and so on up to unnilnillium for the hundredth (fermium; Fermi, do not weep!), and so on, forever more, having thus far perpetrated unnilseptium (107), unnunniloctium (108), and so on. There is much to be said in favor of this procedure, and its real purpose, which I have slightly misrepresented, should be kept in mind. First, the new system is not supposed to replace the familiar names of the kingdom—as if the United States were to be renamed Unland and the United Kingdom Quadquadland, ranked, perhaps, according to the number of telephones. Rather, these systematic names are there to enable explorers to speak of the as-yet-unnamed. If there is indeed an Atlantis to the south, then potential explorers now have the language to spin their yarns about the properties of its regions, without having to anticipate history by guessing that a particular region will one day be amundsenium or scottium. Once the flags are planted on the regions, poetry can displace prose and immortality can be invested.

Much more interesting than the workmanlike usefulness of systematic nomenclature is the patchwork of names that have arisen over the ages. The ancient names, whose origins are lost or only faintly discernible in the mists of history (like those of many European states), include sulfur (from the Sanskrit *sulvere*), iron (Anglo-Saxon *iron*), gold (Anglo-Saxon *gold*), and silver (Anglo-

Saxon *seolfor*). Copper, though known for millennia, takes its name, via the Latin *cuprum,* from Cyprus, where much was found.

When prehistory gives way to the era of history, origins can be assigned with more confidence. We have already seen that Humphry Davy named sodium and potassium after their respective sources, soda and potash. A similar source of names is found just to the east in the Western Rectangle of metals: calcium is present in lime (Latin: *calx*) and magnesium came from ores obtained from a white earth, *magnes carneus*, found in Magnesia, a district in ancient Thessaly. Over in the Eastern Rectangle, nitrogen (from the Greek *nitron* plus *genos*, or "the niter giver") is named in recognition of its occurrence in nitrates. The name of its eastern neighbor, oxygen (from the Greek *oxys*, acidic), is a mistake. In 1777, when Antoine Lavoisier named it, oxygen was widely believed to be a universal constituent of acids. That view has since been overthrown (hydrochloric acid, a compound of hydrogen and chlorine, is a counterexample), but the names of elements, like those of one's children, stick.

Some elements have been named on the basis of their color. Two prominent examples are chlorine, a pale yellow-green gas, from the Greek *chloros* (yellowish-green); and iodine, a violet solid, from the Greek *ioeides* (violet). Colors also figure in the names of other elements, but they are not immediately apparent in the element. For example, rubidium, from the Latin *rubidus* (deep red, or flushed), is not a red element: it is the typical metallic gray of the Western Desert. However, compounds of rubidium, when burned in a flame, give out a ruddy glow. Likewise, cesium renders a flame sky blue, and is named from the Latin for that color (*caesius*). Thallium, from the Greek *thallos* (green shoot), is applied to that element because its compounds turn a flame green-shoot green. Even deeper in the personality of some regions lie the roots of their names. Vanadium of the Isthmus, for instance, forms a series of compounds that spans a rainbow of colors, and this elegant element is named after the Scandinavian goddess of beauty. Colorful chromium, from the Greek *chroma* (color), likewise is given a name that reflects the range and vividness of the colors of its compounds. Iridium, from the Greek and Latin *iris* (rainbow), also received its name for this reason, and likewise rhodium, from the Greek *rhodon* (rose).

Other senses, other names. The nose is a constant companion of a chemical explorer, and it is not surprising that some elements have been given names that are acknowledgments of the nose's response. A pleasant fragrance is rarely an attribute of the elements; where the nose is active, it generally responds with a wrinkle. Bromine, for example, the pungent and vaporous lake of the Eastern Rectangle, takes its name from the Greek word for "stench" (*bromos*). Farther west, in the Isthmus, lies osmium, another smelly element (*osme* is Greek for "smell").

Many regions are named after locations that have a particularly close association with their occurrence. Strontium, named for Strontian, in Scotland, is one example, but there are many more. Continents have been raided to name the regions: Europe has been borrowed for europium, and America, presumably the North, has been borrowed for americium. There is as yet no asium, africium, or australium, and presumably there will never be either arcticium or antarcticium. Numerous countries and their geographical features in one disguise or other are hidden in the names of the kingdom, and a happy winter's evening fireside game is to identify them. Scandium from Scandinavia, francium from France, and germanium from

Germany are easy. More difficult are rhenium from the Rhine (Latin: *Rhenus*) and ruthenium for Russia (Latin: *Ruthenia*) from its discovery in the Urals. States and cities can be spotted in the landscape; they include californium and berkelium, both acknowledging the awesome contribution of the University of California at Berkeley to the extension of the kingdom. Disguise, as I mentioned, has often been applied, and sometimes a city is an etymologically subterranean presence: Hafnium, in the western Isthmus, is a more than flimsy disguise for Copenhagen (Latin: *Hafnia*). Holmium, one of the lanthanides, is perhaps easier to penetrate, giving way under reflection to Stockholm (Latin: *Holmia*). Beneath lutetium lies Paris (Lutetia, City of Light). No place, though is more widely commemorated in the kingdom than the little Swedish town of Ytterby, just outside Stockholm. Ytterby has been stretched, clipped, and otherwise distorted in recognition of its richness as a mineral source, where prospectors of the real world have found numerous regions of the kingdom. Yttrium and the lanthanides ytterbium, terbium, and erbium, though not yet great contributors to the world economy, are nevertheless memorials to this prolific region of the actual globe.

Then there are the people who find their eternal memorial in the kingdom. Lecoq, as we have seen, has had his little joke with gallium. Most people don't presume to give their own names to elements, and the other personal names have all been bestowed by committee in the nominee's honor. It is appropriate that the southern shoreline, including the southern strip of the Southern Island, should commemorate forever—or for as long as we exist—the contributions of Albert Einstein (einsteinium), Enrico Fermi (fermium), Dmitri Mendeleev (mendelevium), Alfred Nobel (nobelium, for stimulation to discovery, maybe, rather than discovery itself), and Ernest Lawrence (lawrencium, for the Berkeley scientist who originated the atom smashers now used worldwide for land reclamation in the kingdom). The most recent names have been attached to the regions of the mainland's southern coast, where atoms that exist for less than a blink of an eye have guaranteed immortality to several scientists (fig. 7). The names ascribed by international committees to these elements—in some cases, contrary to the wishes of the actual discoverers—are still the cause of some annoyance and are by no means universally accepted. Here, currently, lie dubnium (for Dubna, the site of a Soviet contribution to extending the kingdom), joliotium (for Frédéric Joliot-Curie), rutherfordium (for Ernest Rutherford), bohrium (for Niels Bohr), and hahnium (for Otto Hahn). Yes, there are women: curium (for Marie Curie) is found on the southern strip of the Southern Island, and meitnerium (for

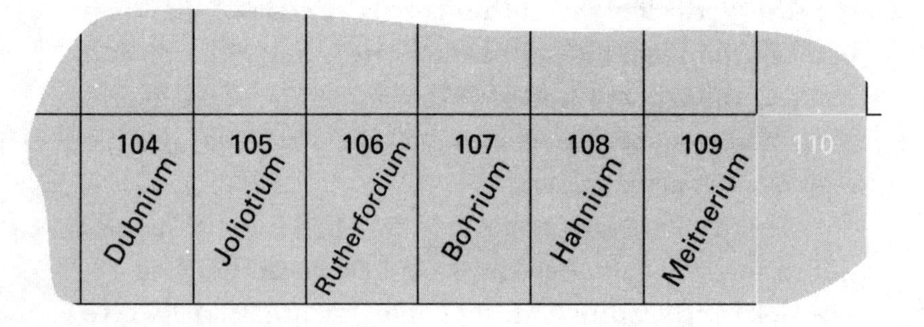

FIG. 7.

A squabble over nomenclature is taking place in this region of the kingdom. Proponents of other names or of permutations of these names have suggested scrapping the names "dubnium" and "joliotium" altogether and renaming elements 104 and 105 rutherfordium and hahnium, respectively; naming element 106 seaborgium in honor of Glenn Seaborg; giving bohrium its full names of nielsbohrium; and renaming element 108 hassium in honor of the German state of Hesse.

Lise Meitner, Hahn's coworker) lies on the south coast next to hahnium, and is currently the last outpost of the kingdom.

Gods as well as mortals have lent their names. The prodigious strength of the titans (the children of Gaea, the Greek Earth goddess) is commemorated in titanium, and Prometheus is remembered in the lanthanide promethium. Fleet-footed Mercury has given his name to fleetly flowing quicksilver (mercury). Evil spirits are also represented, as in nickel and cobalt, both from German words for "goblin" (*Nickel* and *Kobold*), and for the difficulties they caused in the extraction of copper from its ores. The most memorable characteristic of the lanthanide dysprosium seems to have been that it was very difficult to isolate (Greek: *dysprositos*, "hard to get at"). Some elements seem to have been named by mistake: besides oxygen, a misnomer we have already noted, there is the heavy metal molybdenum, from the Greek *molybdos* (lead); and silvery platinum, from *platina*, a diminutive of the Spanish *plata* (silver).

Celestial bodies other than the gods have given their names to many elements. Selenium was named for the moon (Greek: *selene)*, for its silvery appearance. The lanthanide cerium was discovered in 1803, two years after the discovery of the asteroid Ceres, and palladium in 1803, at about the same time as the discovery of the asteroid Pallas. The southern strip of the Southern Island—elements made in war for purposes of war—commemorates the bellicose gods Pluto, Neptune (who, besides being a sea god, was the god of earthquakes), and Thor, the Norse god of thunder. Rather oddly, there is no marsium or venusium. If Atlantis ever rises from the ocean in peacetime, then perhaps sunnier gods and goddesses will give their names and we shall have aphroditium and venusium or their non–Greco-Roman counterparts from other traditions.

CHAPTER 6

THE ORIGIN OF THE LAND

The regions of the kingdom needed to form; they have not been there forever. In this imaginary cosmos, they rained down from above and filled the entire terrain. Some of that elemental rain fell in the twinkling of an eye, as soon as our true universe came into existence, about fifteen billion years ago. At the big bang, the cataclysm that shook spacetime into being and marked the inception of our universe, the offshore Northern Island of hydrogen poked up through the sea of nothing and came into existence. Hydrogen was the very first element to be formed. Only that tiny foothold of the kingdom was formed in that instant of cosmic enormity, but it was a seed from which all the kingdom's awesome richness was to spring.

Almost as soon as the island of hydrogen appeared above the waves, the Northeastern Cape of helium appeared. In the violence of the first three minutes, hydrogen atoms smashed together, such was the tumult and density of the primeval storm, and melded into helium, and the first region of the kingdom's mainland rose above the waves. Seen as a landscape of cosmic abundance, the offshore island of hydrogen is a towering column soaring into the sky, and helium, the Northeastern Cape, is a substantial column a quarter of hydrogen's height.

There the kingdom rested, with these two lone pillars virtually all there was to indicate the riches to come. Minutes passed, then years, thousands then millions of years, and all there was of the kingdom remained the same. There was as yet no hint of empire.

Although its material composition remained constant, other things were happening in the universe. Now that matter had formed—even such primitive and tenuous matter—events occurred that left a tangible imprint, a memory, an inheritance. The great clouds of the two primitive elements that pervaded all there was of space were not completely uniform, and the weak gravitational attraction between atoms slowly led to the formation of regions of greater density and the slow denuding of others. The universe became lumpy. At last, structure was beginning to form, and the ultimate (or, at least, currently the ultimate) descendant of that primordial lumpiness is you and me.

The incipient structure grew, and the differentiation of matter from empty space became more pronounced. Even the lumpiness was not uniform, and within the enormous clouds of greater density there were smaller regions of even greater density. In due course, those smaller regions were to become stars, and the great regions that contained them were to become the galaxies that still hang above us in the sky. While these events took place and portentous lumpiness formed, the Northern Island and the Northeastern Cape stood almost alone, the only two elements that existed in the universe to any significant extent.

The formation of the stars, like the flood of discovery that accompanied our terrestrial advances in technology and the deployment of energy, opened up new opportunities, and soon the surface of the imaginary ocean was broken by the gradual appearance of new regions of the kingdom. These new regions broke through the water in the far northwest, forming a strip of northern coastline stretching east from lithium to beryllium and across to the northern elements of the Eastern Rectangle. Now at last, and forevermore (maybe), the twin pinnacles of the North were not alone.

The source of these new lands—the origin of the great majority of the elements—was the turmoil within the stars. As stars formed, the hydrogen atoms within them were given new opportunities to collide with one another and recaptured the vigor they had at the beginning, a billion years ago. Hydrogen smashed against hydrogen and added a little more helium to the kingdom. The Northern Island lost a little height and the Northeastern Cape grew slightly taller, but the changes were barely perceptible. That erosion and corresponding elevation continues today, and it will persist until the stars go out, for the energy released in this process of *nuclear fusion* is the fuel that lights the stars, including our own star, the sun. Moreover, the process of fusion gave rise to other elements: hydrogen fused with helium, and lithium sprang from the sea. Lithium collided with hydrogen, and helium with helium, and beryllium was born.

In spite of all this stellar turmoil, not much of the primordial hydrogen and helium has been converted into other elements, but what there is of these other forms of matter, though only a slight contamination, is vital to our existence. These other elements have been formed to different extents and are present in different cosmic abundances. To see the unevenness of the terrain of the kingdom when cosmic abundance is plotted against location, we need to descend from the towering heights of hydrogen and helium and inspect the subtle undulations of the much lower-lying remainder of the kingdom. Seen from hydrogen's height, the terrain of the mainland is almost a uniform plain, sloping down from the North, rising toward iron, and then falling away again to regions that barely break the surface along the southern shore. However, when we land at lithium and trek eastward, the land is far from smooth. In fact, there is a pronounced undulation, with lithium, beryllium, and boron marking low ground and carbon, nitrogen, and oxygen, the most abundant elements after hydrogen and helium, forming great heights, with a somewhat lower peak to the south, at iron.

To understand these and other undulations we shall meet on the trek through the kingdom, we need to have the merest whisper of knowledge about the internal structure of atoms. The pebbles of the land are not solid, like real pebbles, but have an inner structure (fig. 8). Indeed, if you were to pick up a pebble from anywhere in the kingdom, you would be surprised at its weight and astonished at its almost gossamerlike lack of being. An atom is in fact seemingly almost nothing. Only a superhumanly sharp eye will catch sight of a microscopic dot at the center of the gossamer—a dot that despite its minute size is responsible for almost all the atom's mass. This massive yet microscopic dot is the *nucleus* of the atom. As far as we are concerned, nuclei are made up of two types of fundamental particle, *protons* and *neutrons*, tightly gripped together (the exception is hydrogen, with a nucleus that consists of a single proton). When we speak of atoms colliding in the interior of stars, we really mean that these minute nuclei are colliding. Nuclear fusion is the coalescence of protons and neutrons to produce a more complex and more massive nucleus, and hence a new element. It is for this reason that the formation of the kingdom is called *nucleosynthesis*.

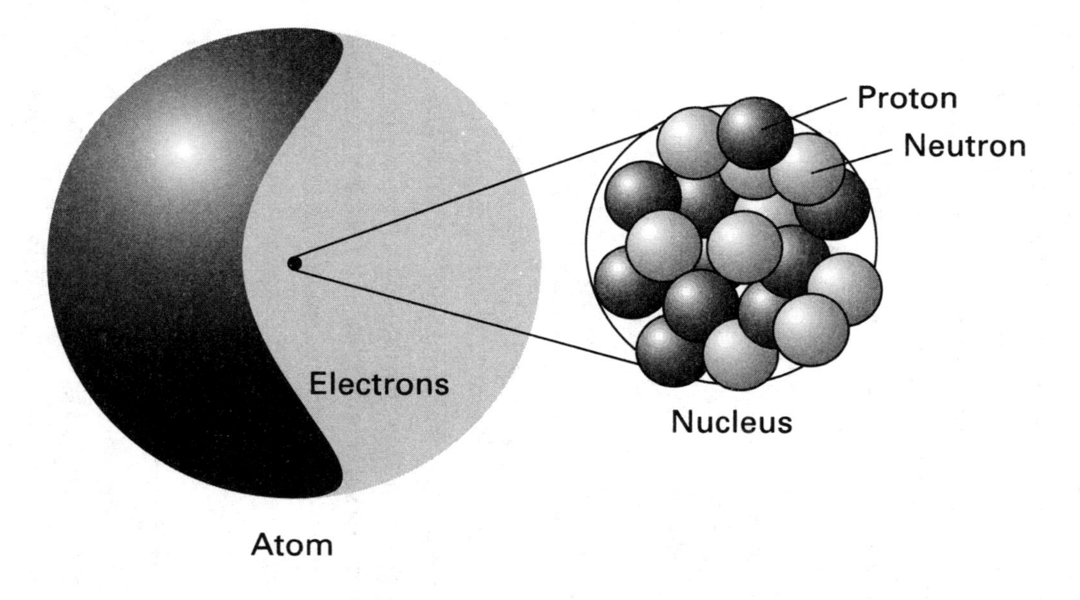

FIG. 8.

The internal constitution of atoms. An atom consists of a cloud of electrons surrounding a minute central nucleus, which is about one-hundred-thousandth the diameter of the atom (and therefore much smaller relatively than displayed here: about the size of a fly at the center of a football stadium).

The stability of nuclei in the hot tumult of the interior of stars is a matter of the greatest importance, for a flimsy nucleus may well be shattered in its next collision, and collisions take place every billionth of a second. Nuclei survive because their protons and neutrons stick together as a result of a special force (prosaically but evocatively named the *strong force*) that these particles exert on one another. Working against this strong force is the repulsion between like charges, characteristic of the positively charged protons squashed together in the nucleus. A nucleus can survive only if there are enough neutrons—electrically neutral particles—to provide an additional source of the strong force without contributing to the repulsion that tends to rip the nucleus asunder. And, specifically, it can survive in the tumult of the stars only if there are enough protons and neutrons to grip each other tightly, rather like survivors on a flimsy raft in an ocean tossed by a storm. In fact, it is surprising that any nuclei can survive in stars. Beryllium (four protons and five neutrons) and boron (five protons and five or six neutrons) *just* manage to survive, but they are far from abundant in the universe, for many were shaken apart almost as soon as they were formed. Much of the lithium, beryllium, and boron we now have are fragments, chipped from larger nuclei.

It is as well for us that beryllium and boron do survive, because they are the doorway through which nucleosynthesis must proceed in order to form the King of Mediocrity, carbon, and thus eventually allow the kingdom to be appreciated by minds. There is in fact something rather special about a nucleus of carbon which enables it to form quite rapidly. If this special feature—its technical term is *resonance*—were absent, there would not be the current abundance of carbon in the universe but only a smidgen of the element. Resonance is the effect that lets one pendulum couple strongly to another pendulum of the same frequency, and by extension is the coupling between a nucleus and a proton of the appropriate energy.

Without that resonance, there would be no life. Because of it, a lot of carbon has formed; it is the third most abundant element in the universe. Moreover, the formation of carbon opens the door to the nucleosynthesis of other elements, and the rise of the regions of the northern shore of the rectangle, the next rank to the South, and lands as far as iron on the Isthmus. However, as we trek through these regions, we notice that although the land is rising toward the low hills of iron, there is a regular alternation. Now the kingdom is displaying another correlation between property and location, for alternate regions are higher than intervening regions: alternate elements are more abundant than those lying between them. Along the northern shore, for instance, carbon, oxygen, and neon are crests and nitrogen and fluorine are troughs of the generally falling terrain. The pattern, with minor modifications, is found throughout the kingdom, as it slopes southward to the sea. It is also found on the Southern Island, which is more like a saw than a smoothly descending surface. It will presumably be characteristic of Atlantis, if Atlantis is ever reached.

The explanation of this rippling of the kingdom is to be found in the details of nuclear structure—that is, in the way that protons and neutrons stack together. If there are even numbers of protons and even numbers of neutrons, then especially stable stacking patterns can form; the nuclei of alternating elements therefore have a marginally enhanced stability compared with their neighbors. But there are two problems with the scenario we have described so far. One is that after iron the nuclei become so large that the protons and neutrons are not able to interact with one another as strongly as is the case in smaller nuclei, and there is a decline in stability. If a star were simply to burn to its death, then the cinder left behind would be iron alone and the kingdom would not extend beyond the first row of the Isthmus. Iron is the end of the road for the process of nucleosynthesis we have described, and without a more varied history there would be less than half a kingdom. The second problem is that there is little to be gained for life in the universe if even half a kingdom is confined to the stars. We need the kingdom to exist outside the stars, where events can transform inorganic matter into organic matter and thence to organisms.

Both problems are solved by an inevitable consequence of nucleosynthesis which it would be helpful to look at now in a little more detail. The first stage in the life of a newly formed star begins when its temperature has risen to about 10 million degrees (10^7 K). This is the hydrogen-burning stage of the star's life cycle, when hydrogen nuclei fuse to form helium. Even in our sun, a middle-aged star, about 600 billion kilograms of hydrogen are processed in this way each second. When about 10 percent of the hydrogen in a star has been consumed, a further contraction takes place, and the central region of the star rises to over 100 million degrees. At the same time, the outer regions are pushed away from the central region by the activity within, and the inflated star becomes what is known as a red giant. Now helium burning can begin in the dense hot core, and helium nuclei fuse into beryllium, carbon, and oxygen. This phase continues until the helium in the core is exhausted and carbon and oxygen are present in approximately equal proportions (it is noteworthy that these two elements are the most abundant in the universe after hydrogen and helium). At this stage, the basic building blocks of life are present.

At the end of the helium-burning phase, the inner regions of the core again contract and the temperature rises once more. In sufficiently massive stars—those with a mass at least four times that of our sun—the temperature can rise to a billion degrees (10^9 K), and carbon burning and oxygen burning can begin. These processes result in the formation of elements that are on the edge of being seriously heavy, including sodium, magnesium, silicon, and sulfur. Now the elements of the future terrestrial landscape have been formed.

When the stellar core has been depleted of carbon and oxygen and is rich in silicon, the silicon-burning phase can begin, and silicon is converted to sulfur, argon, and other heavier elements. If contraction can raise the temperature of the interior to about 3 billion degrees, then the so-called *equilibrium phase* of the star's life cycle begins, and elements close to iron are formed. Iron is the most stable nucleus of all; as noted, if a star were to burn to its end, it would become a ball of iron.

Meanwhile, in the outer regions of the star, nuclei may be subjected to intense streams of neutrons, which have been generated by the nuclear processes in the core. These neutrons are captured by the nuclei in collisions, and several additional neutrons may accumulate in a given nucleus. At some stage of neutron accumulation, the nucleus becomes so unstable that it spits out an electron—a sign that, in effect, a neutron has collapsed into a proton and a new, heavier element has been formed. Thus, the kingdom gradually extends beyond iron, and elements up to uranium (and perhaps beyond) are formed.

Stars do not burn smoothly from their inception to their death. As a star consumes atomic nuclei, there comes a stage at which its center is old and worn, depleted of fuel. At that stage, the outer regions of the star may collapse, like a falling roof, and plunge into the hot abyss below. The falling matter crashes onto the star's dense core and bounces back; in this way, a star may shrug off its outer regions, scattering them through space. The parent star may continue to burn, and may even explode again, but—more to the point—it has scattered precious ash, the newly formed elements, through space. Space no longer contains only tenuous clouds of primordial hydrogen and helium, for now there is an important contamination. The first great pollution has taken place, and the potential for landscape, life, technology, and power now exists outside the stars.

Other stars may form, and now when they do they form from gases already contaminated with the terrain of much of the kingdom. These condensed, contaminated clouds of gas ignite, burn in their nuclear way, and forge a richer soup of the kingdom. In time, it is their turn to die, in the explosive way typical of their kind. When the starquake comes, the splash and countersplash spread even more of the elements through the cosmos.

The newly liberated nuclei scattered through the cosmos have a complex composition that reflects the various modes by which they have been made. Stars of different mass burn in different ways: some never reach equilibrium; some never even reach the helium-burning stage; some burn rapidly, others slowly. However, one common feature is that the mechanism of nucleosynthesis leads to very little lithium, beryllium, and boron, for these light elements are either bypassed in schemes of synthesis or are consumed almost as soon as they have been made. These regions of the kingdom are all but beneath the surface of the sea of nonbeing at this stage in the evolution of matter. But the journey of nuclei through space is a dangerous one, full of cosmic rays, or streams of rapidly moving particles. Collisions take place, and chips are struck off heavier nuclei, a process called *spallation*; these chips include nuclei of lithium, beryllium, and boron. These regions of the kingdom rise from the sea by virtue of processes occurring far from the stars, in the treacherous quiet of interstellar space.

Now the kingdom is virtually fully formed. It rises above the sea of nonbeing and will remain substantially the same almost forever. The kingdom was formed in and among the stars, and all there is around us in the imaginary kingdom and in the actuality of our own Earth was once forged in distant, long-extinguished stars and then scattered to the winds of space in the stellar death throes.

After their birth, the elements had a gentler history, freed from the angry turmoil of stellar interiors. After drifting aimlessly through space, deflected by collisions, impelled by radiation, and swept along by moving clouds of gas, some of these wandering atoms collected in one or another of the clouds. (Note how the history of the kingdom so often includes clouds that slowly acquire a more structured form.) So it is with one particular cloud—a cloud mainly of hydrogen and helium, but now filthy with representatives of the entire kingdom. It is four or five billion years ago, ten billion years after the universe began. The dirty cloud condenses, as dirty clouds do when gravity can draw the particles together despite the vigor of the processes that oppose contraction. Eventually it bursts into nuclear activity, releasing energy as hydrogen fuses into helium, and brightening the surrounding sky. But not all the dirty dust condenses. A crucial remnant remains, encircling the incandescent star, bumping together and sticking together as grains, then rocks, then boulders, then planetesimals, and finally great spheres speeding around a sun—our sun. One of these spheres in due course becomes recognizable as the molten planet that will be the platform of all our activities: Earth.

The point of ranging this far in the history of the kingdom is that there are two landscapes of abundance. So far, our mind's eye has seen the kingdom from the cosmic standpoint, where hydrogen and helium tower above a gently undulating plain. Now we shall view the kingdom with its Earthly abundances plotted. The landscape looks quite different in this light. Gone are the great pinnacles of hydrogen and helium, replaced by little more than their foundations. Now Mounts Iron, Oxygen, Silicon, and Magnesium tower over the terrain, accompanied by the slightly lesser peaks of sulfur, nickel, calcium, and aluminum. What cataclysm can so have transformed the kingdom that where there were towering peaks there are now mere stubs, and where there were stubs there are now peaks?

Although the formation of the Earth was a much gentler affair than the formation of the elements in the stars, it was still vigorous by current standards. In particular, the young Earth was molten throughout, as its insides still are today, and the great heat drove off many volatile compounds. Any elemental hydrogen trapped in bubbles would return to space, and so would helium, being chemically inert and thus unable to anchor itself to other elements. Apart from the hydrogen that managed to bond and form nonvolatile compounds, including the water imprisoned inside minerals, the pinnacles of hydrogen and helium simply boiled away, and we are left with their stumps. Virtually no helium remained on the primitive Earth; the remnant we measure on Earth today is helium that has been created by the radioactive decay of heavy elements such as radium and uranium.

Elements able to form compounds that anchored them despite the heat remained as part of the young Earth. Silicon and its northern and western neighbors oxygen and aluminum formed the silicates and aluminosilicates—the rocks we currently stand on—and these light elements floated on the denser elements, like iron and nickel, that sank to the planet's molten depths, where they still lie. Some elements formed unwise alliances with sulfur—unwise because many sulfur compounds are volatile and were ejected from the boiling Earth. The cosmic landscape of the kingdom's terrestrial abundances rose and fell, according to the volatility of the compounds the regions formed, and the current landscape of the kingdom seen in terms of terrestrial abundances differs greatly from the landscape in the light of the cosmos as a whole.

We have seen a little of the distant past and of the recent past, and we broadly know the present of the kingdom. What, though, of its future? Is the kingdom eternal, or will it slip below the waves? There is a good chance that one day—in a few years, or a few hundred years at most—Atlantis will be found, which will be an intellectual achievement but probably not one of great practical significance. Much more interesting is the fate of the kingdom in the very distant future, long after all the stars have died and all our ink has returned to the skies. A likely (but not certain) scenario is that in that distant time, perhaps 10^{100} years into the future, all matter will have decayed into radiation. It is even possible to imagine the process. Gradually the peaks and dales of the kingdom will slip away and Mount Iron will rise higher, as elements collapse into its lazy, low-energy form. Provided that matter does not decay into radiation first (which is one possibility), the kingdom will become a lonely pinnacle, with iron the only protuberance from the sea of nonbeing. But even iron will decay, and in due course that lone pinnacle will sink below the waves.

It is conceivable that our achievements can be embedded in radiation alone, so there may be an era in which a memorial to our former presence—our information, and our knowledge of the former kingdom—will exist. However, it is also conceivable that there will be an even more distant era, in which all that radiation will have been stretched out into flatness as the universe expands, leaving no imprint. Now the universe will be no more than dead, flat spacetime, and there will not even be an intellectual existence of the kingdom. Now the kingdom is truly below the waves. All that once was, including the memory and the knowledge of what once was, will have vanished.

CHAPTER 7

THE CARTOGRAPHERS

Initially, the fact that the kingdom was rational was not appreciated. Certainly, it was known in the eighteenth century that there were regions for hydrogen, oxygen, iron, and copper, and by early in the nineteenth century several dozen other regions were known. But they were thought to be an archipelago of disconnected islands, scattered at random. Indeed, at the time, there was hardly any reason to believe that these disparate islands were cousins.

The first serious, albeit tentative, cartographer of the kingdom was a German chemist named Johann Döbereiner, born the son of a coachman in 1780 and later apprenticed to apothecaries. As an assistant professor of chemistry at the University of Jena, he noticed that reports of some of the kingdom's islands—reports brought back by their chemical explorers—suggested a brotherhood of sorts between the regions. In 1829, he identified triads in the archipelago, which suggested that some of these islands were at least cousins. Döbereiner's triads were little more than groups of elements that, to the casual eye, manifested a gentle gradation in physical and chemical properties, yet Döbereiner pointed out that the triads had one remarkable feature: the atomic weight of a triad's central member was the arithmetic mean of its two outer members. A numerical basis of consanguinity and location is beginning to emerge. In one region of the archipelago, for instance, he identified iron, cobalt, and nickel (elements 26, 27, and 28, with atomic weights 55.85, 58.93, and 58.69) as a triad, and one zone of the Isthmus appeared. Elsewhere, he linked chlorine, bromine, and iodine (elements 17, 35, and 53, with atomic weights 35.45, 79.90, and 126.9, which occur one after the other in the north-south row of halogens), and calcium, strontium, and barium (elements 20, 38, and 56, with atomic weights 40.08, 87.62, and 137.3, also running north-south).

Spotting triads is not all that difficult, from our point of view, for similarities between neighboring regions are quite close. It is much harder to achieve a global view—over, say, dozens of elements. A further difficulty was that in Döbereiner's day, only a few regions of the kingdom had been reported, and there was only scant information about them; in particular, their atomic weights were only very poorly known. Döbereiner was attempting to identify patterns in a patchwork in which great regions had not yet been sewn into place, using misleading data. It is a testimony to his chemical acumen that he managed to identify the relationships he did. Why the triads related to one another by way of the apparent coincidence of atomic weight was still far from clear: at that stage, the triads seemed to be unrelated to one another. From the vantage point of retrospection, we can see that the triads lie in distinct zones of the kingdom. Döbereiner and his contemporaries saw fragmented and patchy ground, and not the global pattern.

There were certain dangers in proposing relationships. Few chemists in the late eighteenth and early nineteenth centuries believed—indeed, had any reason to believe—that there was an arithmetic pattern to matter. Matter was tangible; numbers were abstract. Abstractions could certainly have patterns, as they were the product of intellect and were invented systematically. Matter, on the other hand, was the stuff of the Earth, not a figment of the intellect. Matter was *real*. Its components might be teased out and identified one by one, but there was no reason to suppose that they fell into a numerical pattern.

Even when patterns were first identified, some expressed their scorn. By the 1860s, a sufficient range of regions had been explored so that it was feasible to ask whether there were extended rhythms of the kingdom. The first global pattern was proposed by the French geologist Béguyer de Chancourtois in 1862, who arranged the elements on a spiral inscribed on a cylinder and managed to accommodate twenty-four elements in this way. He noted a periodicity of properties, with similar elements recurring after every seventh element.

In 1864, a superior, two-dimensional arrangement, in which the origins of the current map of the kingdom are readily discerned and which managed to locate thirty-five elements, was proposed by the English chemist John Newlands. Newlands, who had Italian ancestors, was born in London in 1837. He fought with Garibaldi's army at Naples, and in due course percolated downward, insofar as romantic and heroic activity is concerned, to a career as an industrial chemist in a sugar refinery. Unfortunately, he chose the dubious analogy of music to report his observations. He noticed that, as in a musical scale, harmonies among the elements were seen in every eight steps. Thus, he proposed that elements lay in octaves, and to a certain extent he was correct. From our retrospective altitude, we can see that the northern coastline of the rectangles of the kingdom has eight regions, from lithium across to neon. If for the time being we pretend that the low-lying eastern coast of noble gases does not exist, for it was not above the sea of ignorance in Newland's time, then the seventh element after lithium is not neon but sodium, and this region, another alkali metal, has pronounced similarities to lithium. Another octave of elements, seven steps to the east from sodium (again ignoring the eastern coastal plain), brings us back to potassium, another of sodium's relatives. When he arranged the regions according to the masses of their atoms—that is, their atomic weights—he found that every eighth known element (the noble gases were a problem for the future) created a harmony.

De Chancourtois and Newlands had glimpsed the layout of the kingdom. Until then, apart from Döbereiner's observations of scattered kinship, reports of the regions were as a scattering of independent islands. It was truly unfortunate for Newlands, however, that he chose the analogy of music to express his observations, for he was laughed to scorn. How could the fundamentals of nature be related to musical harmonies? How absurd! Had Mozart picked out chemical combinations when he composed? Had Haydn only concocted potions to pour into and thereby soothe ears? Why did not Mr. Newlands try listing the elements alphabetically instead?

This ridicule, however, did not stifle further attempts to organize the kingdom. Three scientists pursued the quest for rationalization of the seemingly irrational. One was William Odling, Michael Faraday's successor at the Royal Institution and later a professor of chemistry at Oxford. In 1864, the same year in which Newlands published his musical map, Odling published a chart of the kingdom which also bears a close relation to modern charts. He placed fifty-seven regions in locations we would recognize today, with one or two regions lumped together and not properly identified. He also left gaps, which strongly suggested the existence of missing regions. Odling's work has been largely and unjustly ignored. It is of mild interest that he lived to a great age (ninety-one) and alone among the early cartographers survived well into the twentieth century: he died in 1921.

In the same year of 1864, the German chemist Julius Lothar Meyer showed that the ability of elements to form compounds with one another varied periodically with atomic weight, and put forward an abbreviated chart of the kingdom. He also examined the physical properties of the regions, and particularly the volume occupied per atom (the "atomic volume"), which he could calculate from the density and the atomic weight. He found that by plotting atomic volume against atomic weight—at the time the most fundamental parameter for distinguishing the elements and putting them in order—a rhythm emerged (fig. 9). Newlands' octaves were certainly there, for the atomic volume peaked at the eighth and the sixteenth elements. But then the rhythm changed, and the peaks were eighteen elements apart. Following that peak, and insofar as the elements were known, there was another long period before the peak in atomic volume was repeated. Meyer had uncovered a periodicity in the properties of the elements—a complex rhythm that appeared to be a wave passing through their ranks. The kingdom was more complex than John Newlands had guessed: Newlands had discerned only the beginnings of complex structure; Odling and Meyer had penetrated further.

Far away, in St. Petersburg, the bigamist and chemist Dmitri Ivanovitch Mendeleev (1834–1907) had his eye on a

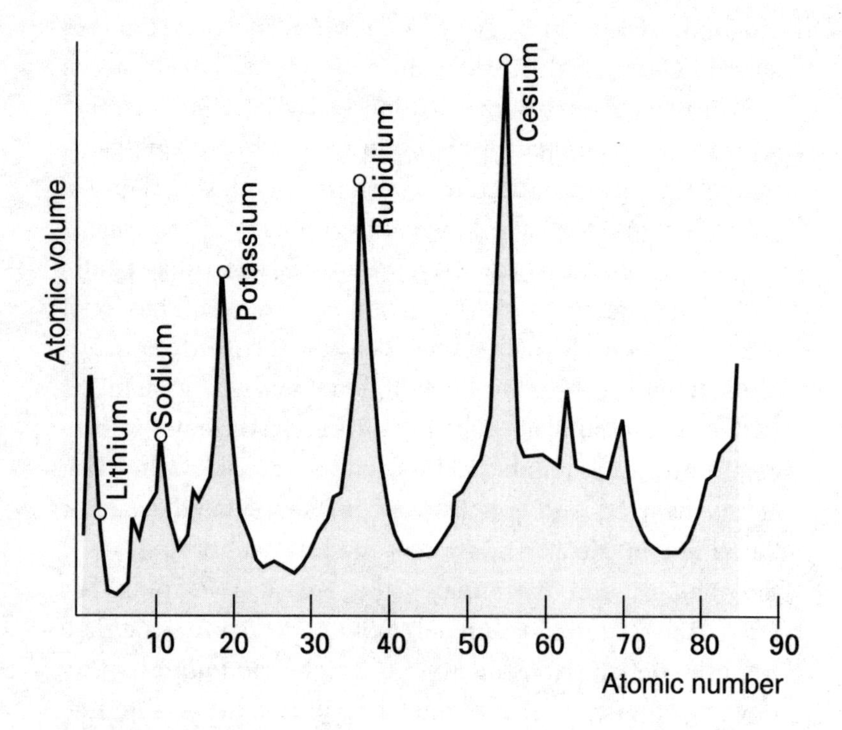

FIG. 9.

A modern (and more complete) version of Julius Lothar Meyer's plot of atomic volume of the elements. He plotted the values against atomic weight; here we have plotted them against atomic number, which is a more fundamental property. Note the periodic rise and fall of the values.

....

similar problem, but was tackling it from a more chemical point of view and working apparently in ignorance of what Odling and Meyer had achieved. Mendeleev looked a little like the mad monk Rasputin and had a reputation to match. He was born in Siberia, the youngest of fourteen children. In St. Petersburg, he proved to be outspoken, radical, and quarrelsome. The charge of bigamy stems from a Russian law then in force that required him to wait for seven years after divorce before remarrying: he did not, marrying immediately. The Czar refused not to receive him at court, on the grounds that although Mendeleev had two wives, he, the Czar, had only one Mendeleev.

The Czar's one Mendeleev knew of sixty-one islands of the kingdom (fig. 10), and sought the chemical relationships between them that would order them into a unity. Only three-fifths of the current kingdom was known; he had to have courage to listen to his inner voice and to conclude that some regions had not yet been discovered. Better to leave gaps to complete later, perhaps to guide

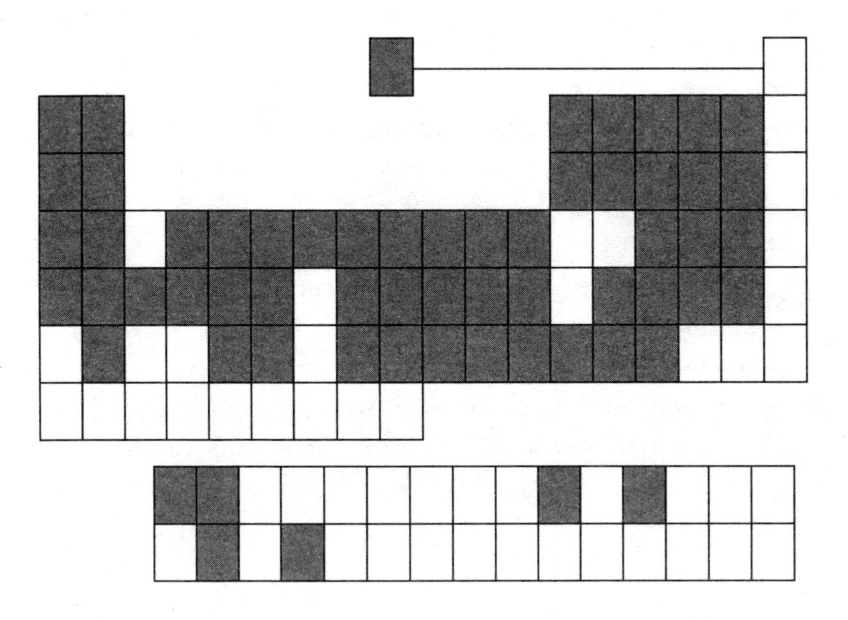

FIG. 10.

The regions shaded are the elements known in Mendeleev's day. You can identify them by comparing this chart with the labeled chart on the endpapers.

prospectors into the kingdom, than to corrupt the pattern by imposing a false appearance of completeness. It is said that during a brief nap in the course of writing a textbook of chemistry, for which he was struggling with the problem of the order in which to introduce the elements, he had a dream. When he awoke, he set out his chart, in virtually its final form. The date was February 17, 1869.* It may be relevant to the process of discovery that Mendeleev enjoyed playing a form of solitaire (patience), and that he had written out the elements on cards, which he dealt into rows and columns.

Mendeleev argued that atomic weight was the only fundamental characteristic of an element, because it was independent of temperature and other variables. He found that by arranging the elements in order of increasing atomic weight, many of which were then known with reasonable accuracy, and having the courage to leave gaps where the pattern seemed to need them, and setting the regions out in a rectangular array that resembles the kingdom's current layout, he captured similarities. First and second cousins lay in columns north to south, just as in the kingdom today, and there is a gradual blend of properties from west to east. Today the vertical columns are called *groups* and the horizontal rows are called *periods*. Mendeleev knew only sixty-one elements, compared with the hundred or so we now know, and he charted only about thirty-two in his first attempt. This was enough to show the lay of the land. He had to tinker with the order here and there, and he permitted his chemical nose to

*This date is according to the old Julian calendar. It corresponds to March 1 on our Gregorian calendar.

guide him, where numerology would have led him astray. Thus, according to an order based on atomic weight alone, he would have had cobalt and nickel switch places, and he would have done the same with tellurium and iodine. He ignored the masses of the atoms in these cases, and used his nose to assign the elements to their appropriate groups: iodine was in all chemical respects a member of the halogens, so he steered that element to its current location. Cobalt most plainly—at least to a chemist's nose—must lie between iron and nickel, not in the order suggested by its atomic weight.

Where gaps were needed to make the regions belong to their appropriate groups, gaps were left. As we have seen, the existence of these gaps pointed explorers to the discovery of unknown lands, and *eka*-silicon (germanium) and *eka*-aluminum (gallium) were soon discovered in the real world and used to plug the gaps. Mendeleev also made some mistakes, proposing the existence of lands that were never to be found.

Mendeleev's cartography of the kingdom was a little different from the maps we use today, but then first attempts at organization are often different from their descendants. For one thing, he did not consider the kingdom to have an Isthmus; the regions were all clustered into a rectangle. Since his justly celebrated map, the style of presentation has evolved, and even in the late twentieth century has not fully settled down.

The modern map of the kingdom has detached itself from the tyranny of atomic weight. The elements are no longer listed by mass, but by a much more fundamental quantity, *atomic number*. The atomic number of an element is the number of protons in the nucleus of one of its atoms; the number of neutrons, which of course contribute to its atomic weight, are ignored. Thus, hydrogen has a single proton, and its atomic number is 1. Helium (which has either one or two neutrons) has two protons, and its atomic number is 2. Uranium has ninety-two protons, and its atomic number is 92. This unambiguous ordinal number rises smoothly through the table: it rises west to east along a row, and each more southerly row has a higher atomic number. The landscape of the kingdom, when viewed in the light of the atomic number, is a smoothly rising surface, a surface without blemish, sloping upward from northwest to southeast. A trek across the kingdom along a period leads to no false steps, as are occasionally found when atomic masses are displayed instead. Even the offshore island in the southern sea rises uniformly on its northern strip, and then again, at greater altitude, along its southern strip. If we were to tow it to the mainland, we would know exactly where to slot it into place.

Of special importance is the fact that because an ordinal number can be attached, by independent measurement, to each region of the kingdom, we know that no element is missing. Had the concept of atomic number been known before the end of the nineteenth century, then it would have been immediately apparent that there was a strip of undiscovered land just east of the halogens, for the numbers would all have jumped by 2 between the eastern end of one period and the western beginning of the next period. It is also for this reason that we can say with confidence that there is no analogous undiscovered strip, or whole prairies of lost worlds, west of the Western Desert. There are simply no missing atomic numbers. Granted that the atomic number *is* fundamental (and there is no one who doubts that it is), then the roll call of the elements is complete—at least, between 1 and 109.

The association of an atomic number with each region of the kingdom also means that we can step along the southern coastline, ticking off each step of the way by determining the atomic number of the reclaimed land. If, as has happened, the atomic number jumps by 2, then we know that a region has been missed, and we can leave a gap. The atomic number is not assigned on this dangerous, fleeting southern coast by order of discovery: it is assigned by specific measurement, and the region is assigned its location accordingly. For a similar reason, and from our knowledge of nuclear structure, we know the approximate distance offshore of at least one of the regions of the Island of Stability: it lies at an atomic number close to 115. We can even speculate on the size of the Island of Stability, for theories of nuclear structure suggest that there is another region at atomic numbers close to 180.

The other great advantage of having ordinal numbers ascribed to the regions of the kingdom is that there can no longer be squabbling over the location of a region (although we shall see that other opportunities for territorial squabbles remain). Indeed, as is invariably the case with all modern maps, the erstwhile ambiguities of the locations of wayward regions—cobalt and nickel, tellurium and iodine—all vanish: these elements turn up where explorers with chemically attuned noses expect to find them. We have to conclude that it is largely a happy accident that atomic masses increase in step, in most cases, with atomic number, for that tendency enabled the early cartographers to grasp a correlated atomic property even though it was not the fundamental property that later investigations provided.

As for the squabbles, there are three; two chemical, one bureaucratic. The first chemical squabble (a clubby squabble, not a World War) centers on the Southern Island. Chemists know exactly the order of the elements in this region of the kingdom, and they know where the island lies when it is towed ashore and slotted into the mainland. The order of the regions is fully determined by their atomic numbers. However, the decision about the extent of the kingdom that should be excised from the mainland and towed offshore depends on judgments about which elements are sufficiently similar to form a close alliance with one another. There is room for disagreement at each end of the island, particularly with regard to the rather complicated relationships between the regions of the island's eastern end. Some maps tow slightly different strips out to sea. In our map of the kingdom, we have towed away the greatest strip of the mainland, a superset of all Southern Islands.

Why, though, tow any regions away at all? The reason is purely pragmatic: the map with the kingdom complete is simply too long and skinny to sit conveniently on a page. (Indeed, the Isthmus of the transitional Western Desert is sometimes discarded for this reason, when a point is being made about general trends of behavior across the kingdom. We shall not use this hypertruncated form, but you may well come across such a map and be puzzled by an apparently collapsed kingdom.)

The second chemical squabble is in the far North, and concerns the location of the offshore Northern Island of hydrogen. To those who do not like offshore islands, there is the problem of where to put it on the mainland. This is the war of the Big-endians versus the Little-endians. Big-endians fight to tow the island ashore to form a new Northwestern Cape, immediately north of lithium and beryllium and across from the Northeastern Cape of helium. There are good chemical reasons for doing this, and good reasons (as we shall see) based on the internal structures of these atoms, which are all closely related structurally. But it is also a little awkward, because hydrogen is a gas and not a metal, like all the other regions of the Western Desert, and its location there would thus be incongruous. The Little-endians would tow the island to shore in the opposite direction, and attach it to the mainland north of the halogens and next to helium, doubling the extent of the Northeastern Cape. Hydrogen, they argue, is certainly a gas, and it has a chemical and structural resemblance to the halogens. Nevertheless, the arguments for including it there are hardly more compelling than the choice of the Big-endians, and on internal structural grounds (as we shall see) its location there is incongruous. Some chemists compromise, by allowing a region marked hydrogen to appear at both locations. We shall take a different line, which is based on an observation made long ago—that the elements of the northern shore, the starter-fringe of the kingdom, do show a number of striking differences from their more southerly neighbors. It is therefore reasonable to suppose that the very first element of all is, to a corresponding degree, a significant stranger to the land, and that it would be inappropriate to tow it to shore anywhere, for that would suggest a false similarity. So, in our map of the kingdom (but not in some others you may meet), the region of hydrogen is firmly anchored off the northern coast.

The bureaucratic squabble concerns the labeling of the groups. The periods (the horizontal rows) are quite straightforward, and no battle rages here: hydrogen and helium make up Period 1, and successive horizontal rows are labeled Periods 2, 3, and so on. The offshore Southern Island is considered to be a part of the mainland for this exercise: its northern strip is a part of Period 6 and its southern strip is a part of Period 7. Land reclamation is currently going on in the east of Period 7. The numbering of the groups (the vertical columns) is where the battle is taking place. The following remarks may well seem confusing, because they *are* confusing. Mendeleev and his immediate successors labeled the groups of the Western Rectangle I and II, and those of the Eastern Rectangle III through VIII. They also labeled the columns of the Isthmus I through VIII. Group VIII of the Isthmus, being particularly pronounced in its transitional character and unlike any of the other groups, consisted of three elements in each row. For instance, iron, cobalt, and nickel—a classic Döbereiner triad—were all ascribed to Group VIII. Copper, which comes immediately after them, was assigned to Group I of the Isthmus, a notation that it had to share awkwardly with the distant alkali metals (there are some faintly compelling chemical reasons for this sharing). Similarly, the group to which zinc belonged was forced to share the label Group II with such elements as magnesium and calcium in the Western Rectangle. To distinguish between the Rectangle groups and the Isthmus groups, the letters A and B were used. Thus the alkali metals were assigned to Group IA and copper and its cohorts to Group IB; magnesium was in Group IIA and zinc (a cousin about ten times removed) was in Group IIB. However, there were Cavaliers and Roundheads as well as Big- and Little-endians afoot, and whereas Cavaliers used this nomenclature, Roundheads reversed A and B. Thus, according to the Roundheads, copper lay in Group IA and zinc in Group IIA. The Döbereiner transitional triads, according to the Roundheads, were in Group VIIIA, and the noble gases were in Group VIIIB; while according to the Cavaliers the triad was in Group VIIIB and the gases in Group VIIIA. There was even a sect that broke away to call the column occupied by the noble gases, down on the seashore of nonreactivity, Group 0—for which there is no Roman numeral.

It will be understood that here there is plenty of opportunity for confusion, and, since familiarity breeds entrenchment, for bloody resistance to any hint of change. To sweep aside this international confusion, the human overseers of the kingdom formed a kind of United Nations to agree on the nomenclature they would use.* They came up with a sensible suggestion: sweep away the Old Order and replace it by the New. So, instead of using the Roman numerals I to VIII, with the sectarian 0 and the alphabetical sores of A and B, we now simply use the Arabic numbers 1 to 18 for successive columns of the mainland, starting in the far West with 1 and ending at the far East with 18. The columns of the Southern Island are anonymous in this scheme, but that should cause no hearts to bleed; as these regions are almost indistinguishable chemically from one another they have no great need for distinguishing labels. According to the kingdom's United Nations, the regions of the Isthmus all now occupy individual groups, numbered from 3 to 12.

Our map of the kingdom uses this current scheme, but the Old Guard still—for good reasons, which will emerge when we study the institutions of the kingdom—clings to the old, believing that it captures more directly the rhythms of chemistry, which is, after all, the principal point of the existence of the kingdom. Nevertheless, the New Guard appears to be gaining ground, and although the battle is far from won and is raging gently on paper, with skirmishes here and there and the sound of cannon rolling across the Western Desert, current signs are that the New Guard will ultimately win but will show respect for the Old Guard by slipping sometimes into its usage.

---

*IUPAC, the International Union of Pure and Applied Chemistry, the committee that also oversees the naming of new elements.

It is appropriate to bring up one final point of nomenclature at this point. We have referred to the Eastern and Western Rectangles of the kingdom, the Isthmus joining the two, and the offshore Southern Island. It will come as no surprise to learn that these zones of the kingdom have names of a more official character. Each blocklike zone is known, prosaically, as a *block*. So, there are four main blocks of the kingdom. Each block has its distinguishing label. Until we have studied the institutions and government of the kingdom, these labels will seem arcane, for they are rooted in a technicality that we shall shortly unfold. For reasons that will at this stage seem utterly obscure, the Western Rectangle is called the *s-block*, the Eastern Rectangle is called the *p-block*, the Isthmus is called the *d-block*, and the Southern Island is called the *f-block*.

That, broadly speaking, is where the cartography of the kingdom currently lies. In summary, the regions are arranged in increasing value of atomic number, and lie in horizontal periods (numbered 1 to 7) and vertical groups (numbered 1 to 18) that form a series of four blocks, labeled *s*, *p*, *d*, and *f*.

However, before tunneling deeper into the organization of the kingdom, we need to emphasize that cartography is not a static science. There have been many attempts to map the kingdom in more elaborate ways, particularly (as in the depiction of the Earth itself) by drawing it in three dimensions. Is the kingdom, like the Earth, a globe? Can relations be shown more clearly and accurately in more dimensions than two, as in the depiction of the Earth? Would a higher dimensional representation suggest a more fruitful, more accurate projection than the one we have described?

It is worth taking a moment to enter the cartographers' lair to see what proposals have been made. Most are long-forgotten museum pieces that have had little influence on navigation of the kingdom. We have already seen that de Chancourtois envisaged the kingdom as a spiral on a cylinder: this was his *telluric screw*, arcanely so-called because the region of tellurium (named, appropriately enough, from the Latin word for "earth") lay at the central point. This spiral and its more complex successors, which have included pretzellike re-entrant spirals and double spirals, neatly avoid the impression that there is a discontinuity in the kingdom, since West is contiguous with the East, much as Alaska meets Siberia.

But because this is an imaginary kingdom, there is no need to emulate the Earthlike solid figure of true kingdoms. One of the most interesting versions of the Periodic Kingdom is a weathervane model (fig. 11). In this model, the *s*-block is the axle (note that it has two faces), and the *p*-block (which also has two faces) sprouts out from it between Groups 2 and 1 and forms a vane; in turn, the two-faced *d*-block sprouts out of the line where the *s*- and *p*-blocks meet to form another vane, and the two-faced *f*-block sprouts out from the *d*-block, to form a vane on a vane on a vane. Thus, all the zones of the kingdom are treated similarly, and if a *g*-block, say, were ever to be discovered, it would sit comfortably as yet another vane. There are numerous variations on this theme, but all suffer from the same difficulty as our conventional globes: they take on more than two dimensions.

There may well turn out to be a deeper portrayal of the kingdom than any that has been discovered yet, and the power of visual display provided by computers may one day give us richer, more suggestive displays of the relationships of the kingdom than can be had from any flat map. An entirely different foundation for the portrayal may well be devised, or layers of portrayal, which will enrich our understanding beyond our current imagination.

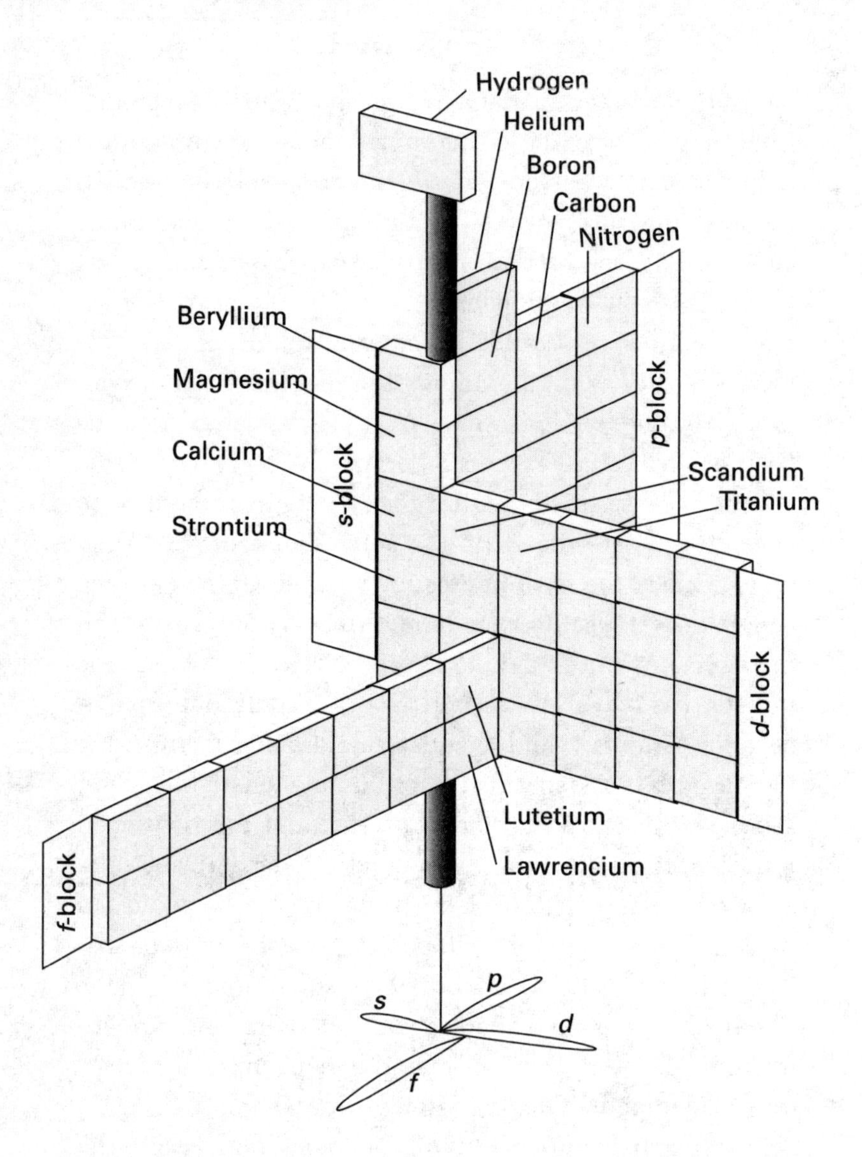

FIG. 11.

One of the three-dimensional versions of the chart of the kingdom. This four-vaned version was suggested by the chemist Paul Giguère. Each vane is a block. Only one side of each block is visible in this view. As a challenge, I suggest you try to identify the regions: some names have been added to start you off.

For the rest of this account of the kingdom, however, we shall return to Earth and use the contemporary, well-thumbed chart of the kingdom that has been in our mind's eye all along and sprang from Döbereiner's triads, Newlands' octaves, Odling's table, Meyer's rhythms, and Mendeleev's global insight.

PART 3

GOVERNMENT AND INSTITUTIONS

CHAPTER 8

THE LAWS OF THE INTERIOR

The laws of the interior, the rules that govern the structures of atoms and result in their characteristic properties, require a certain suspension of belief. Well, that may be putting the matter a little too strongly, for the laws that govern the kingdom of the elements are far simpler than those that govern our own lives. It could even be argued that they are far simpler than those that govern the behavior of billiard balls, planets, and other macroscopic objects. Quantum mechanics, that symbol of the break with the past that is so characteristic of twentieth-century science, is essential for understanding the structure of the kingdom, accounting for its rhythms, and seeing why the regions lie in a particular juxtaposition. In fact, quantum mechanics is central to the interpretation of the kingdom. It is inevitable that we speak of it if we are to convert an empirical chart of family relationships into a more useful and understandable device.

Quantum mechanics is necessary because it is inevitable that we speak of atoms, and atoms, being microscopic, can be discussed only in its language. According to this description of nature, energy can be transferred to an object only in discrete amounts called *quanta*, not smoothly and continuously, as was supposed in classical physics. Furthermore, the distinctions made in the macroscopic world between particles and waves fade, and the two concepts merge into one. In the quantum world, entities are regarded as having the attributes both of particles and of waves, and according to the kind of observation being made, one or the other description may be more appropriate. The kingdom is ruled by this discreteness and wave-particle duality, and to understand its institutions and administration we need to accept these seemingly bizarre ideas.

We have already seen that a pebble of the kingdom, by which we mean an atom of an element, is a gossamer entity, more space than substance. To be specific, it consists of a massive but minute central nucleus surrounded by almost empty space. That not completely empty space is pervaded by the most important (for chemistry) of nature's fundamental particles, the electron. So our first crude picture of an atom of an element is a speck of a nucleus surrounded by an electron cloud.

The existence of atoms was first established experimentally at the start of the nineteenth century, when the Manchester schoolteacher John Dalton carefully analyzed the masses of substances that combined with one another. Dalton had no direct evidence for the existence of atoms, but he inferred from his measurements that some unchangeable entity was involved in chemical reactions. Today we have much more direct evidence for atoms, and our instruments—essentially elaborate refinements of microscopes—can now display them for all to see.

The internal structure of atoms was determined by a succession of experiments in the late nineteenth and early twentieth centuries. That electrons are a universal constituent of matter was established by J. J. Thomson, working at Cambridge University. He showed, by using an apparatus that was the forerunner of the modern television cathode-ray tube, that a certain kind of fundamental particle could be stripped out of any element. George Stoney named this fundamental particle the *electron*, and in due course its mass and its negative charge were determined. The entire kingdom, in a sense, is carpeted with electrons, and we shall see that the personalities of the elements are a consequence of the ways in which these electrons are arranged; here (in helium) they are grouped in twos, there (in magnesium) a dozen are clustered together. Michael Faraday would have been delighted at this discovery, for he, the archwizard of electricity, was deeply confident that electricity was bound up in some way with the composition of matter. After all, electrolysis had shown that matter could be wrought into different forms by the passage of an electric current. Through Thomson's work it became clear that electrons were universal, that matter somehow was built from them, and that an electric current was nothing more than a stream of electrons in transit.

When electrons were first discovered, there were two profound problems concerning their contribution to the structure of atoms. One problem was that it was not known how many electrons were present in a given atom. Because an electron was known to be thousands of times lighter than an atom, it was thought that even hydrogen, the simplest element, might contain hundreds of electrons; we know now that there is only one electron in this cosmically Paleolithic element. The second problem was the nature of the atom's *positive* electric charge, which must be present for the atom as a whole to be electrically neutral. Was it, as Thomson believed, a jellylike background of positive charge, in which hundreds (or thousands) of electrons were embedded, or was it something rather more structured, more sophisticated?

Nature being nature, it turned out to be more sophisticated. In 1910, at Manchester University, Ernest Rutherford turned his formidable experimental talents toward the resolution of the structure of the pebbles of the kingdom. Working under his supervision, his students Hans Geiger and Edward Marsden shot alpha particles—tiny, positively charged particles formed by the radioactive decay of heavy elements—at a thin sheet of gold foil. While most of the particles sailed right through the foil, some were scattered by it, and a few were reflected right back to the source. Having expected blobs of jelly to pass through blobs of jelly, like blancmanges hurled at a wall of blancmanges, Rutherford later mused that "it was quite the most incredible event that has ever happened to me in my life," and that the experimental results were as surprising as if "a fifteen-inch shell fired at a piece of tissue paper came back and hit you." These blobs, if blobs they were, had an extraordinary reflective property. Closer study of the results led him to the view that an atom was not a spherical blob of soft jelly at all, but that its positive charge was concentrated into a massive central speck, and that most of the rest of the atom was empty space. He announced his results in 1911, and thus the *nuclear atom* was born. In this model, the atom has a single, central, specklike, massive nucleus, which contains the atom's positive charge and almost all its mass, and a surrounding tenuous atmosphere of electrons, with enough electrons present to cancel exactly the positive charge of the nucleus. Each pebble of the kingdom is this nuclear entity, a speck electrically commanding almost empty space.

Rutherford was able to estimate the positive charge of

gold nuclei and obtained the equivalent of a few dozen units of charge. Therefore, he concluded, there were not thousands of electrons in an atom, but a few dozen at most. This number was measured, in a different type of observation, by another of Rutherford's students, the young physicist Henry Moseley, shortly before he was cut down by a sniper's bullet and added his blood to the earth of Gallipoli. By investigating the properties of X rays emitted by atoms, Moseley was able to count the positive charges of a nucleus. In this way he was able to determine that central organizing ordinal number, the *atomic number* of an element, then regarded as the number of units of positive charge on its nuclei. Hydrogen, as we have seen, has atomic number 1: so its nucleus has a single unit of positive charge and is surrounded by a single electron to cancel the nuclear charge. A carbon atom has atomic number 6, with six units of positive charge on its nucleus and six electrons around the nucleus to form the neutral atom. The pebbles of the kingdom's southern shore have atomic numbers close to 100; so they have close to a hundred electrons each.

At about this time it became possible to determine experimentally the masses of nuclei accurately and precisely. At last, after more than a century, John Dalton's view that all atoms of a given element were the same could finally be put to the test. It turned out to be, to collective surprise, a test that Dalton's speculation failed. The masses of the atoms of an element were found to have a spread of values. Because all these atoms, regardless of their mass, belonged to the same element, they were termed *isotopes*, from the Greek for "equal place." Some elements, particularly the light ones, consisted of one or at most two or three isotopes, but the heavier elements in the southern part of the kingdom were often composed of up to about a dozen or so isotopes. Clearly, and as we have already suspected, the mass of an atom is not a fundamental quality of an element of the kingdom.

The realization that an atom of an element has a unique and characteristic atomic number yet, within narrow limits, a variable mass prompted the view that the atomic nucleus itself had an inner constitution—that it was made up of smaller entities. A model emerged in which each atomic nucleus consists of a group of positively charged fundamental particles called *protons*, equal in number to the atomic number of the element. Thus, all atoms of hydrogen have a nucleus that contains one proton, all atoms of carbon have a nucleus that contains six protons, and all atoms of uranium have a nucleus that contains ninety-two protons. These numbers are invariable: change the number, and you change the element. They are the unique and qualifying discriminant of the element. However, it was also found that each nucleus could contain a variable number of *neutrons*, fundamental particles identical to the proton but lacking its electric charge.

The number of neutrons in a nucleus does not affect the identity of the element but does affect its mass. Typically, the number of neutrons in a nucleus is similar to (and generally somewhat more than) the number of protons, but numbers of neutrons can vary by a few either way—hence the isotopes. Carbon, for instance, almost always has six neutrons in addition to its six protons, but isotopes with seven and eight protons are also known. The range of numbers of neutrons is greater in the south of the kingdom, as the proportion of neutrons needed to help bind these proton-rich nuclei together increases. At uranium, for instance, the 92 protons are accompanied by about 150 neutrons, with 146 being the most common value.

We can now see why atomic weight broadly correlates with the periodicity of the kingdom, and also why it sometimes fails. First, we need to note that many properties of an element are determined by the number and arrangement of the electrons surrounding the nucleus, for these are easily rearranged and some can be stripped away. Because the electric charge of an atom is zero, the number of its electrons must be equal to the number of protons in its nucleus. We can therefore speculate that there is a link between the properties of an element and its atomic number. But the number of neutrons in a nucleus increases in step with, if a little in advance of, the number of protons. It follows that as the number of electrons increases, so too does the total number of protons and neutrons, which is what determines the mass of an atom. Hence there is a correlation between an element's atomic mass and its properties (its number of electrons).

A point of significant detail, however, is that the outcome of a determination of atomic weight, except in special cases, is not the mass of a single atom. Instead, we measure the *average* mass of a sample, which normally contains several different isotopes. There is no guarantee that this average mass will increase exactly in step with the atomic number, and so we can expect slight faults in the landscape of mass. The correlation of properties and atomic weights is therefore imprecise, just as Mendeleev was forced to concede. He had presumed that some values had been determined incorrectly. But we, from our post-nuclear altitude, can see the cause of the deviations and explain them.

CHAPTER 9

THE LAWS OF THE EXTERIOR

We must now turn our attention to the much larger region outside the nucleus—the region inhabited by the electrons. This is where the chemical action lies, and it is here that the reasons for the differences and similarities between the elements of the kingdom are to be found. The compelling and enduring visual picture that many still have in mind when they think of atoms is of tiny planet-like electrons orbiting a central sunlike nucleus—a picture suggested by the Japanese physicist Hantaro Nagaoka in 1904. This was an early stage in the development of atomic theory, and with the rise of quantum mechanics this picture has been displaced by an entirely different concept. Because the electron has wavelike properties, this newer description of the behavior of matter denies the possibility of an electron's having a precise path. The model we are about to describe was formulated by the Austrian physicist Erwin Schrödinger in 1926, and apart from some refinements it is still the accepted model.

In the current picture of a hydrogen atom—an atom with one electron and with a nucleus consisting of one proton—the electron is distributed as a spherical cloud around the nucleus. The density of this cloud can be interpreted as a depiction of where it is likely for the electron to be found. The cloud is densest at the nucleus, and thins out in regions farther from the nucleus. The most likely place to find the electron, therefore, is at the nucleus itself. This cloudlike distribution of an electron is called an *atomic orbital.* The name "orbital" is chosen to convey a less precise sense than the "orbit" of a planet; however, the imprecision should not be misinterpreted. The density of the cloud at any point can be calculated exactly, and so the form of the orbital is exactly known. The imprecision lies in the interpretation of the orbital in terms of the *location* of the electron. We cannot be confident that an electron will actually be found at a particular point; all we can do is state (and we can do so exactly) the *probability* that it will be found there.

Some people claim that this lack of precision is an indication that we cannot have complete knowledge about any object. A more appropriate interpretation is that the classical picture misled us into thinking we could have more detailed knowledge than is in fact possible. It is better, in my view, to think of quantum mechanics as displaying what truly can be known in the kingdom, and of classical mechanics as misrepresenting what can be known by offering too much in its advertisements.

Hydrogen is a model for all the atoms of the kingdom, but we need several more items of information before we can move on. First, we need to know that a spherical orbital of hydrogen is called an *s-orbital.* Although it is convenient to think of *s* as standing for "spherical," that is only a coincidence, and its actual origin is deep in the history of spectroscopy, where it stands for "sharp," a comment on the appearance of certain spectral lines.

Next, we need to know that orbitals can have a wide variety of different shapes. There are even many different kinds of spherical *s*-orbitals. Thus, if enough energy is pumped into a hydrogen atom, the electron can be puffed up into a second sort of *s*-orbital. If the first orbital is called a 1*s*-orbital, then it is natural to call this one a 2*s*-orbital. When an electron has a 2*s* distribution, we say that it *occupies* a 2*s*-orbital. With even more energy, the electron can be puffed up to occupy a 3*s*-orbital—in which there is a central cloud and two concentric cloud shells—and on into higher *ns*-orbitals (fig. 12).

Now we arrive at a complication, one of several that it will be essential to master if we want to understand the structure of the kingdom. When the electron is given enough energy to occupy a 2*s*-orbital, it can also, instead, form a cloud of an entirely different shape—a two-lobed *p*-orbital. There are three such two-lobed orbitals, according to whether the lobes lie along an imaginary *x*-, *y*-, or *z*-axis—this gives us $2p_x$-, $2p_y$-, and $2p_z$-orbitals, respectively. An interesting and seemingly trivial feature of a *p*-orbital, but a feature on which the structure of the kingdom will later be seen to hinge, is that the electron will never be found on the imaginary plane passing through the nucleus and dividing the two lobes of the orbital. This plane is called a *nodal plane*. An *s*-orbital does not have such a nodal plane, and the electron it describes may be found at the nucleus (fig. 13). Every *p*-orbital has a nodal plane of this kind, and therefore an electron that occupies a *p*-orbital will never be found at the nucleus. We shall see that from such minor differences spring mighty kingdoms.

When enough energy has been pumped into a hydrogen atom, its electron can occupy a 3*s*-orbital. With this amount of energy, it can also occupy any of three 3*p*-orbitals, which are essentially puffed-up versions of the 2*p*-orbitals just described, or it can take up even more complex distributions. Now the freedom it gets from its heightened energy also enables it to adopt any of the five

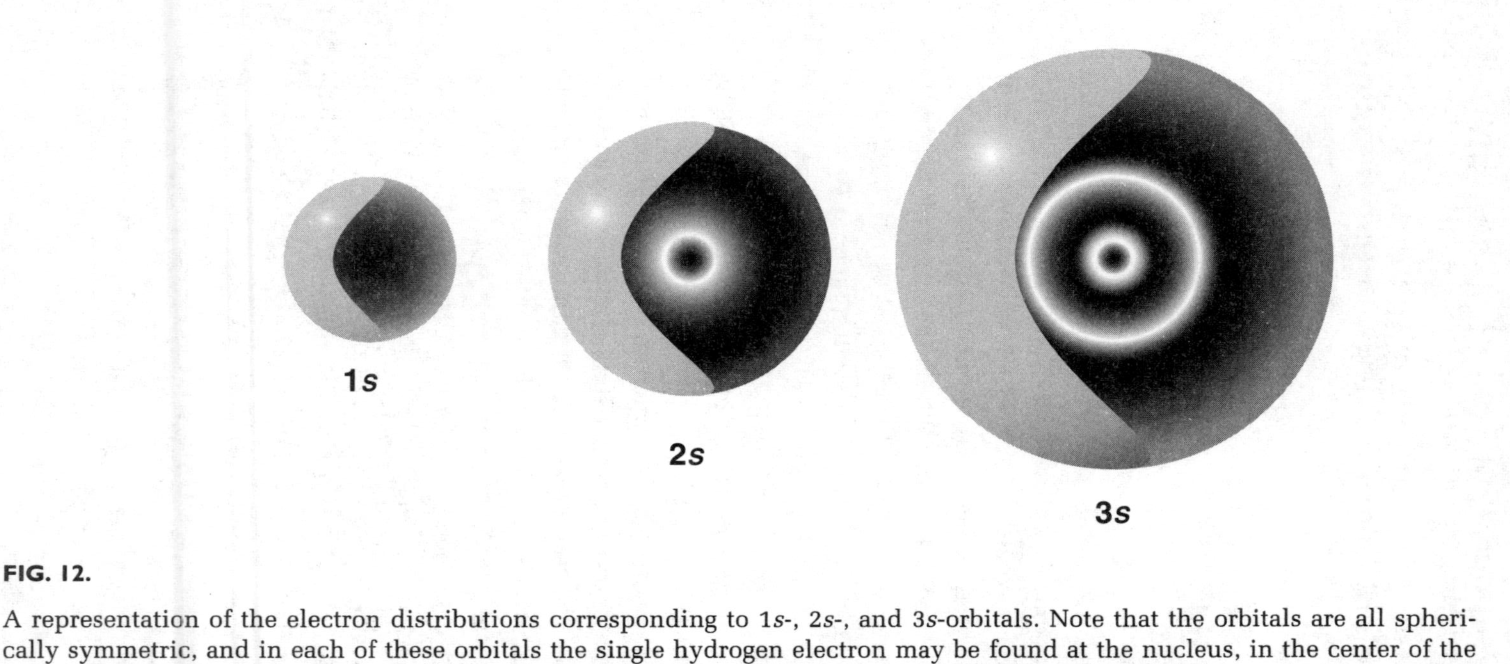

FIG. 12.

A representation of the electron distributions corresponding to 1*s*-, 2*s*-, and 3*s*-orbitals. Note that the orbitals are all spherically symmetric, and in each of these orbitals the single hydrogen electron may be found at the nucleus, in the center of the central cloud. The density of shading indicates the positions where the electron is likely to be found on any plane cutting through the nucleus.

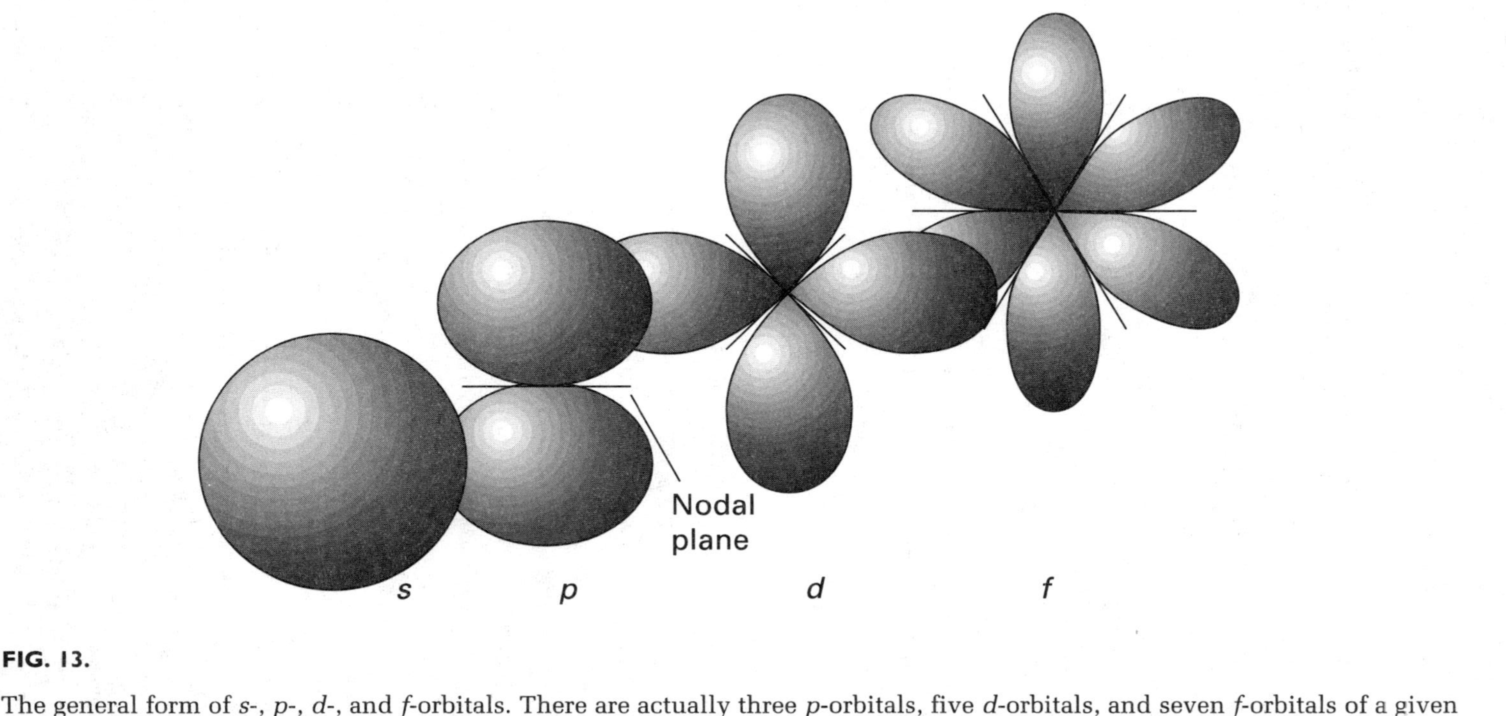

FIG. 13.

The general form of *s*-, *p*-, *d*-, and *f*-orbitals. There are actually three *p*-orbitals, five *d*-orbitals, and seven *f*-orbitals of a given rank of energy, but only one representative of each kind is shown here. The two lobes of a *p*-orbital are on either side of an imaginary plane cutting through the nucleus; there are two such planes in a *d*-orbital and three in an *f*-orbital.

four-lobed cloudlike distributions called *d*-orbitals. Why there are five such orbitals is difficult to justify pictorially, but five there are.

The pattern of possibilities for the distribution of an electron in a hydrogen atom should now be coming clear. In its lowest energy state, which is called the *ground state* of the atom, an electron has the spherical cloudlike distribution characteristic of a 1*s*-orbital. With more energy, the electron distribution can puff up into either a bigger *s*-orbital—that is, a 2*s*-orbital—or one of the three 2*p*-orbitals. With even more energy, the electron distribution can occupy a 3*s*-orbital, one of three 3*p*-orbitals, or one of five 3*d*-orbitals. Even more energy allows the electron to adopt the distribution of a 4*s*-orbital, one of three 4*p*-orbitals, one of five 4*d*-orbitals, or one of the seven six-lobed distributions called *f*-orbitals.* The story goes on, but we need not take it further at this stage.

It might be puzzling why we have dwelt so long on the excited states of hydrogen, when almost every hydrogen atom in the universe will be found in its ground state. What on earth can the structure of the kingdom—with its 109 different regions, only one of which is hydrogen—have to do with states of hydrogen in which even that atom is hardly ever found? That the time is in fact being well spent can be seen by reflecting for a moment on certain letters and certain numbers.

First, consider the letters: the orbitals available to an electron in a hydrogen atom are called *s*, *p*, *d*, and *f*, according to the number of lobes they have. And these letters are also the official names of the Western Rectan-

---

*We have already indicated that *s* is an archaic spectroscopic term meaning "sharp." The terms *p*, *d*, and *f* once likewise denoted features in spectroscopy that were "principal," "diffuse," and "fine."

gle, the Eastern Rectangle, the Isthmus, and the Southern Island, respectively. It is no coincidence that these letters are used in this way, and before long we shall see the connection between the energetic never-never land (or at least hardly-ever land) of hydrogen and the layout of the kingdom.

Next, consider the numbers. There is only one orbital available to an electron in the ground state, and four available in the first higher-energy state. Now we see a relation beginning to emerge: there are two elements in Period 1 (hydrogen and helium), and eight elements in Period 2 (from lithium to neon), which is twice the number of orbitals we have just mentioned. Indeed, if we consider only the Eastern and Western Rectangles, there are eight elements in each row. Can it be a coincidence that the number of columns in the *s*-block is exactly twice the number of *s*-orbitals of a given rank of energy, and the number of columns of the *p*-block is twice the number of *p*-orbitals of any given energy rank? Take the *d*-block. There are typically ten elements in the Isthmus for each period of the kingdom, which is exactly twice the number of *d*-orbitals of any given rank. There seems to be something funny going on in the *f*-block, the Southern Island, where there are fifteen elements in each row, rather than the fourteen that by now the existence of seven *f*-orbitals of a given rank of energy might lead us to expect. However, we have already pointed out that there are wars of occupation raging in this zone, and the division of territory between the island and the mainland is in dispute. The total number of regions across the kingdom in Period 6, which spans the disputed territories, is thirty-two. That number is twice the sum $1 + 3 + 5 + 7 = 16$ for the total number of orbitals, so even if the distribution is in dispute, the overall number is strongly correlated to the number of orbitals available.

At this point, we appear to be on the track of identifying the government of the kingdom—at least insofar as the layout of its administrative units, the arrangement of blocks, is concerned. But what can the excited states of hydrogen have to do with the kingdom? Here we pull the strands of argument together, introduce one new extraordinary law, and reveal the skein of connection.

Following a scheme called the *building-up principle*, first enunciated in 1913 by the Danish physicist Niels Bohr, chemists envisage building the kingdom by the addition of electrons one by one to the orbitals characteristic of a hydrogen atom. Helium, for instance, with its two electrons, is built by putting the first electron into a hydrogenlike 1*s*-orbital and then allowing it to be joined by a second electron. If we denote the structure of a hydrogen atom by $1s^1$, where the superscript denotes the number of electrons occupying the orbital, then helium would be denoted $1s^2$. The 1*s*-orbitals are not exactly the same in both atoms, however, because the more highly charged nucleus of helium draws the surrounding electron cloud closer, but the general shape is the same, and thus the notation is meaningful.

Now consider lithium, an atom with three electrons. This element will teach us two great facts about the administration of the kingdom. First, we need to know that the third electron is not able to join the first two in the 1*s*-orbital. There is a fundamental law of the kingdom that, in an echo of Noah, requires electrons to enter orbitals two by two *and no more than* two by two. That is, according to the *exclusion principle* enunciated by the Austrian-born physicist Wolfgang Pauli in 1924, no more than two electrons can occupy any one orbital. This is an extraordinarily deep principle of quantum mechanics: it can be traced to foundations embedded in the structure of spacetime, and is perhaps the deepest of all principles governing the imaginary kingdom, and hence—because the kingdom is not really all that imaginary—the actual kingdom, too. There is no picture to elucidate the principle: it is handed down on stone tablets as an axiom, from whatever hand carves axioms.

It follows from the exclusion principle that the structure of lithium cannot be $1s^3$, for that arrangement of electrons, with three in one orbital, is forbidden. The third electron must enter an orbital of higher energy than that occupied by the other two. Immediately we see why our excursion into hydrogen's higher energy states is relevant, for the exclusion principle forces additional electrons to enter these higher-energy orbitals. The question we are confronted with, though, is which of the four possible orbitals of the second rank in energy (the 2*s*-orbital and the three double-lobed 2*p*-orbitals) does the third electron occupy? In hydrogen itself, all four orbitals have the same energy, so there would be nothing to choose between them. However, in atoms with more than one electron, the four orbitals do not all have the same energy.

This is the point where the explanation of the kingdom hangs on the existence of nodal planes. Recall that we have said that an electron in any of the *s*-orbitals can be found at the nucleus but an electron in any other type of orbital cannot. With that in mind, try to imagine the distribution of an electron in a 2*s*-orbital in lithium. The electron is mostly distributed as a shell of cloud around the two electrons that occupy the inner, spherical 1*s*-orbital. These two inner electrons effectively cancel two units of positive charge of the nucleus. As a result, the outermost electron, which we shall call a *2s-electron*, experiences only the pull of this partly canceled, or *shielded*, charge. However, the 2*s*-orbital is a cloud with a central core (recall fig. 12), which means that an electron that occupies it can penetrate with nonzero probability right through to the nucleus and thus experience its full attraction. So, as far as a 2*s*-electron is concerned, the nucleus is only partly shielded: penetration partly overcomes shielding.

Now imagine a 2*p*-electron instead (an electron that occupies a 2*p*-orbital). Such an electron is banished from the nucleus on account of the existence of the nodal plane. This electron is more completely shielded from the pull of the nucleus, and so it is not gripped as tightly. In other words, because of the interplay of shielding and penetration, a 2*s*-orbital has a lower energy (an electron in it is gripped more tightly) than a 2*p*-orbital. Thus, the third and final electron of lithium enters the 2*s*-orbital, and its overall structure is $1s^2 2s^1$.

The lesson of the Northwestern Cape of the kingdom is worth repeating, for it teaches us much. After the lesson is absorbed, the structure of the rest of the kingdom becomes immediately explicable (well, nearly immediately) in terms of the building-up principle. First, the occupation of orbitals is governed by Pauli's exclusion principle—that no more than two electrons can occupy any given orbital. Second, the conspiracy of shielding and penetration ensure that the 2*s*-orbital is somewhat lower in energy than the *p*-orbitals of the same rank. By extension, where other types of orbitals are possible, *ns*- and *np*-orbitals both lie lower in energy than *nd*-orbitals, and *nd*-orbitals in turn have a lower energy than *nf*-orbitals. An *s*-orbital has no nodal plane, and electrons can be found at the nucleus. A *p*-orbital has one plane, and the electron is excluded from the nucleus. A *d*-orbital has two intersecting planes, and the exclusion of the electron is greater. An *f*-orbital has three planes, and the exclusion is correspondingly greater still.

We are now sufficiently equipped to walk with understanding through the kingdom. We are already as far as lithium, with its single 2*s*-electron around an inner cloud of two 1*s*-electrons. We shall denote this inner, heliumlike core as [Helium]—or, more briefly, [He]—so the electronic structure of lithium is then denoted $[He]2s^1$, which emphasizes that it has a single electron surrounding an inner, noble-gaslike core. In the construction of beryllium (four electrons), the fourth electron can join the single electron in the 2*s*-orbital, and hence the ground state of this atom—its lowest energy state, which atoms most commonly occupy—is expected to be $[He]2s^2$. The 2*s*-orbital is now full, and we are also at the eastern edge of the Western Rectangle. The next electron, to make up boron's complement of five electrons, will enter one of the three 2*p*-electrons that have been waiting in the wings at slightly higher energy. Thus, boron's ground state is $[He]2s^22p^1$.

Two points should be noted here. The minor one is that it is immaterial which of the three 2*p*-orbitals is occupied: they all have the same energy, and differ only in the orientation of their lobes in space. The major point is that with the start of the occupation of a *p*-orbital, we find ourselves on the western edge of the Eastern Rectangle, the commencement of the *p*-block! Now the origin of the names of the regions is becoming clear: they are the identities of the orbitals that are in the process of being filled by the addition of successive electrons.

There are six columns in the Eastern Rectangle. There are three 2*p*-orbitals, each one of which can accommodate two electrons. As we trek east, from boron to carbon to nitrogen to oxygen to fluorine, we drop one more electron into the three 2*p*-orbitals, and then we arrive at neon. Here we add the sixth *p*-electron, so the ground-state structure of this atom is $[He]2s^22p^6$. Now the 2*p*-orbitals are full, and, indeed, we are at the eastern edge of the Eastern Rectangle. For later simplicity, we shall denote this neonlike arrangement of electrons as [Ne].

It is worth a pause at this point in our journey for your courier to point out two relevant sights. One concerns the numbering of the groups—that is, the chart's vertical columns. It will be recalled from chapter 7 that the groups of the Eastern and Western Rectangles are sometimes numbered from I to VIII. We can now see that these numbers are exactly the total number of electrons in the outermost rank of orbitals. Thus, lithium (Group I) has one electron outside its inner core, beryllium (Group II) has two, boron (Group III) has a total of three (two *s* and one *p*), and so on to the far east, where neon (Group VIII) has eight. This indication of the total number of electrons in the outer regions of the atom is the most compelling reason for sticking to the older notation for the groups. The second point concerns the numbering of the periods (the horizontal rows), which, as we have seen, begins at 1 in the north and increases as you go south. The number of the period is the number of the rank of orbitals in the outermost regions of the atom. Thus, in Period 1 (hydrogen and helium), the 1*s*-orbitals are being occupied. In Period 2, as we have seen, it is the 2*s*- and 2*p*-orbitals that are outermost. In successive rows, we can now anticipate, it is the 3*s*-, the 4*s*-, and so on, orbitals that are being occupied. The period numbers are therefore numbers that identify rank in the concentric shell-like structure of the atom.

We now resume our journey. The eleventh element is sodium; it has eleven electrons, one more than neon. Neon's 2*s*- and 2*p*-orbitals, though, are full, so the additional electron cannot join them. Indeed, there are no more orbitals of this rank remaining, and the additional electron is forced, by the exclusion principle, to occupy an orbital of a higher rank in energy. So it enters the next available orbital, which (as a consequence of penetration and shielding) is a $3s$-orbital. The ground-state structure of the sodium atom is therefore expected to be [Ne]$3s^1$.

Your courier positively insists on stopping the bus at this point in the journey. There is a very special feature of the kingdom suddenly in sight. The addition of one electron has brought us to a structure that is an exact analog of an atom of a region we passed earlier on the journey, lithium. Lithium has the structure [He]$2s^1$, a single s-electron outside a noble-gaslike core; sodium has the structure [Ne]$3s^1$, a single s-electron outside a noble-gaslike core. That is, we have encountered a periodicity of structure: analogous electronic structures of the atoms of the kingdom recur periodically. Indeed, we can now comprehend that sodium lies in the same group as lithium *because their electronic structures are analogous.*

We can begin to sense that the entire basis of the kingdom, the representation of the periodicity of properties, is a manifestation of a deeper rhythm, the periodicity of electronic structure. As the bus is allowed to edge along this awesome ridge of understanding, we can see the truth of this conception unfolding like an extraordinary vista below us. We move on to region 12, magnesium. Its ground-state structure is [Ne]$3s^2$, the expected analog of beryllium ([He]$2s^2$), immediately to its north. Element 13 is aluminum; the additional electron it requires must be accommodated in a $3p$-orbital, because the $3s$-orbital is now complete. Thus, we expect to find aluminum at the western edge of the Eastern Rectangle, and with its structure ([Ne]$3s^2 3p^1$) lying directly south of boron ([He]$2s^2 2p^1$). And there it is! This pattern persists across the Eastern Rectangle, and at argon ([Ne]$3s^2 3p^6$) the s- and p-orbitals of the third rank are full, and we are once more on the coastal plain of the noble gases.

There are few views in the kingdom more awesome than the one we have just seen. The kingdom was established as a portrayal of empirical trends, the recognition of family relationships, the formation of alliances. It was formulated, as we have so far only partially seen, as a summary of much of chemistry. Now, with an astonishing meagerness of concepts—atomic orbitals, the exclusion principle, and the consequences of penetration and shielding—we have arrived at the kingdom's rationalization. It is a considerable achievement to account with such a tiny bag of concepts for the observations made by Döbereiner, Newlands, Odling, Meyer, and Mendeleev.

Your courier must also point out some other, lesser features of the landscape—a vocabulary rather than a vista. We need to know the following language. What we have so far called a *rank* of orbitals is in fact called a *shell.* Thus, the 1*s*-orbital (the only orbital of that rank) itself forms the first shell of an atom, the 2*s*- and 2*p*-orbitals of the second rank jointly form the second shell, and the 3*s*-, 3*p*-, and 3*d*-orbitals jointly form the third shell. These shells can be imagined to be like layers of an onion, with the second shell wrapped around the first, the third around the second, and so on. The groups of orbitals of the same type (the same number of lobes) within a shell are said to constitute a *subshell* of that shell. Thus, the 2*s*-orbital constitutes one subshell of the second shell of an atom, and the three 2*p*-orbitals jointly constitute another subshell of the same shell. When a subshell has its full complement of electrons (two for an *s*-subshell, six for a *p*-subshell, ten for a *d*-subshell, and fourteen for an *f*-subshell), we say that that subshell is *complete.* When the *s*- and *p*-subshells of a given shell are both complete, we say that the shell itself is complete. The *d*- and *f*-subshells are treated slightly differently from their more dominant cousins the *s*- and *p*- subshells, inasmuch as, by convention, they do not need to be complete for the shell itself to be classified as complete. Finally, the outermost shell of the atom, the shell containing the orbitals occupied by the last electrons that built the atom, is called the *valence shell*, and the inner electrons all constitute the *core*. The term *valence* will occur again when we discuss the bonds that atoms form, and its use for the outermost electrons reflects the fact that these electrons dominate the formation of chemical bonds.

We now have sufficient language to travel more articulately from rectangle to rectangle across the Isthmus. Our journey resumes at the end of Period 3, at the noble gas argon, where the third shell is complete. This electron arrangement we denote [Ar]. One more proton in the nucleus, and hence one more electron outside it, swings us back to the distant west, at potassium, with structure $[Ar]4s^1$, in line with the structures of the more northerly elements of this western fringe. Next comes calcium, $[Ar]4s^2$, the southerly cousin of beryllium and magnesium. But at this point there is a change in the rhythm of the kingdom.

At long last, the 3*d*-orbitals come into line for occupation, and instead of our bus leaping to the Eastern Rectangle, we find ourselves bumping over the western abutments of the Isthmus. Now the five 3*d*-orbitals are being occupied, and we bump from scandium ($[Ar]4s^2 3d^1$) to titanium ($[Ar]4s^2 3d^2$), and so on across to zinc ($[Ar]4s^2 3d^{10}$). At this point, on the eastern edge of the Isthmus, the 3*d*-subshell is completely full, and any further electrons are forced to occupy the 4*p*-orbitals, which are next in line for occupation. Now we are in the Eastern Rectangle (the *p*-block) again, and the familiar regions of this rectangle are underfoot until they terminate with krypton. We should note that this first long period, the first period of the Isthmus, comes into existence quite naturally. It is a sign of strength in science that a new phenomenon does not need a different principle for its elucidation.

The rhythm of the southern half of the kingdom is now established, and the next row, Period 5, repeats the structure of the one we have just described, with the 4*d*-orbitals coming into line for occupation. Deep below the kingdom, the exclusion principle and the nodal planes are calling the tune to which, far above in the land of senses, the elements are dancing.

In Period 6, though, yet another rhythm comes to our ears. Now, at last, six-lobed *f*-orbitals are available, and the Southern Island rises from the sea. Fourteen electrons complete the slightly ambiguous northern strip of the island. The Isthmus now resumes, with the *f*-subshell complete, and it, too, soon comes to an end. We have reached the dangerous southern reaches of the Eastern Rectangle, and the period is complete at radon. Another row, another period, and another strip of the Southern Island before the Isthmus resumes. Now, though, the Isthmus does not reach its end, for the regions to the east still lie below the seas. Our journey here is ended, with the kingdom rationalized and the empirical cartographers justified.

CHAPTER 10

REGIONAL ADMINISTRATION

Our early surveys of the kingdom revealed that certain properties of the regions show trends from north to south and from east to west: landscapes viewed in the light of various properties slope from the Northwestern Cape down to the far southeast, or in other systematic directions. Sometimes we saw ridges and sometimes valleys, but there was not a sense of randomness. Even the varieties of regions seemed to cluster systematically: the metals in the expansive Western Desert, the nonmetals in the Eastern Rectangle; the high ridges of reactivity of the alkali metals in the distant west and of the halogens in the distant East. The kingdom is a systematic place, and through our realization that it is a representation of the periodicity of the electronic structures of the atoms, we are in a position to understand the regional variation of its administration, in the sense of seeing how the properties of the elements vary from place to place.

We saw first that the masses of the atoms vary almost systematically throughout the kingdom (recall fig. 3). This variation we have already explained in terms of the rise in the numbers of protons and neutrons in the nuclei of the atoms as the atomic number increases from 1 at hydrogen to 109 on the southern shore. The occasional fault in a perfect landscape is accounted for because atomic weight is calculated as an *average* mass of a sample, which consists of mixtures of isotopes rather than a single type of atom with a unique mass.

Then we saw that the landscape of atomic diameter fell from west to east across a period and increased from north to south, with a peculiar hesitation in the increase inland of the southern shore (recall fig. 4). The general rise in the land from north to south can now be seen to stem from the fact that successive periods of the kingdom correspond to the formation of a new electron shell for the atom. Lithium, for instance, has one electron forming an almost hollow shell around a heliumlike core; sodium, the next element south, has one electron forming a similar shell but now encasing a neonlike core which itself has a heliumlike core, and so on down the group. The electron shells of atoms of successive periods are like successive layers of an onion, and the atoms swell accordingly.

The shrinkage from west to east across a period is not quite as simple to explain, for at first it seems peculiar that atoms shrink as the number of electrons increases. However, we must never forget the role of the nucleus. As we travel east across a period, the nuclear charge increases; this increasing nuclear charge attracts the surrounding electrons more strongly with each successive element, drawing them inward. Although this contraction is opposed by the repulsion between electrons, which increases as their number increases, the contraction generally (that is, almost everywhere) beats the tendency to expand, and therefore the landscape of diameters slopes as you move east. Here and there, bucking the general trend, the landscape rises before falling away again; this occurs toward the eastern end of the Isthmus and in one or two other places. In these locations, the competition is won by the expansion due to the increasing electron-electron repulsion.

There is an important philosophical point here that must not go unremarked. We see that the sloping landscape of atomic diameter, with its occasional gentle rises, is the outcome of a *competition* of finely balanced forces. The effect of nuclear attraction only just dominates the landscape: were it a little weaker, the kingdom would tilt the other way. This determination of landscape by almost equally balanced forces is characteristic of the landscape of the kingdom in all kinds of lights—and it is characteristic of chemistry as a whole. That is why chemistry is such a subtle subject, one where observations are so difficult to predict, for it is difficult to assess whether one particular effect or another will dominate. The kingdom is like a parliamentary democracy with almost equal party representation: sometimes the left will win, sometimes the right.

We still have to explain the peculiar leveling of the plain in the landscape of diameters in the South. You should note that it occurs in the regions that would follow the elements of the Southern Island, if they were towed back to their docking place in the mainland to form a narrow secondary isthmus. This is a clue that it is the presence of these *f*-block elements, and hence the *f*-electrons, that are responsible for the leveling of the plain. The explanation, at least in part, is that the nuclear charge is steadily increasing as we step from west to east along this secondary isthmus, and so we can expect the atoms to contract. Moreover, the *f*-electrons are distributed like clouds of a very spindly shape, and they do not effectively shield the electrons from the increasing charge of the nucleus. Thus, we can expect the effect of nuclear attraction to be dominant, and for there to be a substantial contraction of the atoms across the *f*-block. When the true Isthmus (the *d*-block) resumes, the atoms are much smaller than would be expected if this contraction had not taken place. Indeed, it is found that their diameters are similar to those in the row above, despite the considerable increase in the atomic number, the number of electrons, and the mass. When the elements of the Southern Island are towed back out to sea again, we are left with a leveling of the plain.

Now we see that the landscape of density of the Western Desert of metals—which, as we saw in figure 5, reaches Himalayan heights toward the southern coast, particularly at iridium and osmium, and lingers on to lead—is a consequence of the intrusion of *f*-orbitals into the filling scheme for building atoms of these elements. Thus it is that such a nearly abstract concept as an *f*-orbital can bubble up with consequences for the real world. When next you strain to pick up lead, think of its density as stemming from these near-abstractions and their governance of the kingdom.

The similarity in diameters of the atoms of the Isthmus constitutes another bubble that rises from the subterranean regions of explanation into the real world of technology and commerce. It is largely because of this similarity of diameter that the elements of the Isthmus can mingle to produce the materials we call alloys. Just as a grocer could, if artistry required, quite easily stack apples and oranges in the same pile but would find it difficult to stack oranges and melons together, so a metallurgist can mingle the atoms of the regions of the Isthmus, mixing chromium, manganese, nickel, or cobalt into iron to form the designer metals of the current age. We should realize that it is the *d*-

electrons, the abstractions deep beneath the surface of the kingdom, that are permitting our technology to flourish.

Next, our attention turned to the ease with which electrons could be torn from atoms to form the positively charged particles known as cations. The landscape of ionization energy has a more subtle rhythm than the landscape of diameter, but broadly speaking it is low in the southwest and rises to the Northeastern Cape of helium (recall fig. 6). Even more broadly speaking, it is low in the Western Desert and high in the nonmetal regions of the Eastern Rectangle. This landscape slopes in the opposite direction, in general, from the landscape of diameter. That is exactly what we would expect if nuclear attraction was dominant. As we travel from southwest to northeast, atoms become smaller, and the outermost electron is closer to its nucleus. The grip of the nucleus on the electron is tighter, and so the ionization energy rises. Over in the general direction of the Western Desert, ionization energies are small enough for these atoms to lose their electrons readily. As a result, these elements form metallic solids, in which an array of cations is stacked like oranges in a display and glued together by a pervading sea of electrons. Such ready electron loss cannot take place in the northeast of the Eastern Rectangle, where atoms are small and electrons are tightly gripped, so these regions are not metals.

We can enrich our view of the landscape of ionization energies in a number of ways by drawing on the features of electronic structure that we have at our command. First, we already know that second ionization energies (the energy required to remove a second electron) are larger than first ionization energies because the second electron has to be dragged away from an already positively charged ion. However, there is a pattern to this increase that reflects the electronic structures of the atoms. Take sodium, for instance: it has a single electron outside a compact, noble-gaslike core (its structure is [Ne]$3s^1$). The first electron is quite easy to remove (its removal requires an investment of 5.1 eV), but removal of the second, which has to come from the core that lies close to the nucleus, requires an enormous energy—nearly ten times as much, in fact (47.3 eV). Electron extraction from this core can be achieved (in the sun, for instance, almost all the electrons are blasted off the atoms), but on the whole the energies characteristic of chemical processes are too puny to remove all but the first electron.

Now step east to sodium's neighbor magnesium, with two electrons outside the core ([Ne]$3s^2$). One electron comes off reasonably easily (the first ionization energy is 7.6 eV). The loss of that electron still leaves an electron in an orbital outside the core, and it is still distant from the nucleus and subject to only a weak grip. Thus, although it requires a substantial increase in energy to remove that second electron, the investment required is only 15.0 eV, which is far less than that required to break open sodium's core. This additional energy can be achieved by normal chemical processes. The removal of a third electron, however, requires the extraction of an electron from the core, and the ionization energy for this step is a massive 80.1 eV—the work of physics, not the relatively puny investments of energy typical of chemistry. So, when magnesium forms cations, we can expect it to form doubly charged cations, whereas sodium will form only singly charged cations.

Similar arguments should apply to all the elements of the Western Rectangle, and so we have uncovered another rhythm of the kingdom: the elements in Group 1 should form singly charged cations, and the elements of Group 2 should form doubly charged cations. This is what chemists observe, and almost all the chemical properties of this block of elements can be explained in these terms.

There are not many metals in the Eastern Rectangle (the *p*-block) of the kingdom, but we should expect analogous arguments to apply. One of the commercially most important metallic regions of this rectangle is aluminum. This region lies in Group 13 (old Group III), and has the configuration [Ne]$3s^2 3p^1$, with three electrons outside a neonlike core. We can now predict that this element is likely to lose three electrons moderately readily (within the scope of chemical investments of energy, that is), but no more. This supposition is supported by the first three ionization energies, which are 6.0 eV, 18.8 eV, and 28.4 eV, followed by a leap to an enormous 120 eV for the removal of a fourth electron. Indeed, chemists know that almost all the reactions in which aluminum takes part result in the loss of three electrons, never more and hardly ever less. Incidentally, here we see another good reason that some chemists prefer old labels: the 3 of Group III reveals that aluminum will form triply charged cations, whereas in the label "Group 13" this information is just a little buried.

The difference between this part of the kingdom and the regions of the Western Rectangle is that removal of electrons takes place from two types of orbitals. In aluminum, two of the electrons come from *s*-orbitals and one comes from a *p*-orbital. We might suspect that here is another pattern that careful inspection of the kingdom's soil will bring to light. Indeed, chemical prospectors who travel south down Group 13 will find a change in the disposition of the regions to form cations. Elements in the deep South generally lose just the *p*-electron in their chemical reactions, producing cations that are only singly charged. In the Eastern Rectangle, we have to be aware that some of the regions lose all their valence electrons and thus form cations of a triple charge, while others lose only their outermost *p*-electron, producing a cation with a single charge. Chemists make use of the dual character of these elements to ring the changes on the compounds of these regions.

The variability of electron loss in the Eastern Rectangle is not a peculiarity of Group 13. Next door, in Group 14, lies lead, with the electronic structure $[Xe]6s^2 6p^2$: four electrons (the 4 of 14 in the group label; more obviously in the Roman IV) outside a xenonlike core. We now know enough about the regulations of the kingdom to suspect that lead can lose up to four electrons reasonably easily, and that electron loss beyond that requires breaking into the core. However, we can also suspect, with our developing chemist's eye, that the electrons will come off in two cohorts. We can expect certain reactions to result in the loss of only the 6*p*-electrons, the most weakly bound of the outer shell. In reactions with greater leverage, we can expect to remove all four outermost electrons. Thus, we can expect the chemical properties of lead to be characterized by doubly charged and quadruply charged cations. And so, generally, they are.

We have deliberately not commented on the Isthmus yet. Here in the *d*-block, with electronic structures like $[Ar]4s^2 3d^5$ (that of manganese), the energy of chemical reactions is sufficient to remove the two *s*-electrons and a variable number of *d*-electrons. As a result, all manner of different types of cation can be formed. A major beneficiary of these relaxed circumstances are the processes of life, and also the technological superstructure that life erects from the soils of the kingdom. For example, iron appears as four bloody eyes in the center of the hemoglobin molecule, and its readiness to switch its complement of electrons enables it to accommodate an oxygen molecule and carry it through our bodies to where it is required, depositing it with equal nonchalance, for the greater good. Where electrons are needed to drive a process forward in a living cell, it is often an iron atom that is called on to give one up. In the photosynthetic process, the wellspring of usable energy on Earth, it is manganese that deploys the electrons that the sun's energy releases and that will eventually drive the formulation of our opinions and the execution of our actions.

The chemical industry draws intensely from the Isthmus, because almost all its products are manufactured using catalysts. Catalysts are substances that increase the rates of certain desirable reactions—and, indeed, make some feasible that in their absence could not be carried out. These reactions are almost all based on the electronic complaisance of the metals of the Isthmus. Nitrogen is fixed, formulated as fertilizer, and then used to feed us and rebuild our proteins, through the agency of iron and the readiness with which it lets go of and then retrieves electrons. Bacteria discovered how to make use of molybdenum to fix nitrogen with the investment of evolutionary time rather than financial capital. Sulfuric acid, an essential product of the *p*-block, is made with platinum and vanadium as catalysts, and nitric acid is made with rhodium as a catalyst. Hydrocarbons pumped from carbon's lairs beneath the Earth are cut, chopped, mixed, linked, twisted, extended, clipped, and grown on all manner of special surfaces and compounds acting as catalysts, all of them from the Isthmus. All these deep foundations of society, from the animation of organisms to the viability of industry, are at root making use of the reality of the abstract concept of the *d*-orbital, deep below in the engine rooms of the kingdom.

In this survey of cation formation, there is one last point to make, which concerns their diameters. When a cation is formed, as we have seen, often all the electrons from the atom's outer shell are stripped away to reveal its core. We should therefore expect the diameter of these cations to be significantly less than that of the parent atom, and this is so. Moreover, the variation in diameters of cations mirrors the variation in the diameters of their parent atoms: the landscape falls, with occasional mounds and valleys, from the southwest of the Western Desert to the northeast.

And what of anions? Here the substructure of the kingdom is less variable, and anion formation is confined, under the conditions provided by chemical reactions, to fewer regions. As we saw earlier, the energy of anion formation—that is, the creation of an electrically negative ion by electron attachment—is measured by the electron affinity of an element. Elements with high electron affinities release large amounts of energy when an electron attaches to the atom. Moreover, we also saw that the elements with the highest electron affinities are those lying close to fluorine, in the far northeast of the kingdom. Why is this, and what rhythms of periodicity are running in this part of the land?

It needs to be recalled that an atom of fluorine has the structure $[He]2s^22p^5$, which is only one electron short of the complete valence shell characteristic of neon. When one electron attaches to a fluorine atom, it enters and completes the outermost shell of the atom. The atom puffs up a little, because of the increased electron-electron repulsion, so we expect the fluorine anion (more properly, the *fluoride ion*) to be a little bigger than the parent atom. There is a release of energy of 3.4 eV when the electron enters the atom, because the electron can get reasonably close to the reasonably highly charged nucleus and be attracted to it. Now consider the task of adding a second electron to form a doubly negatively charged fluoride ion. There are two effects militating against the success of this operation: one is that the second electron must be forced on a particle that already bears a negative charge, and like charges repel; the other is that when the electron finally arrives at the ion, suffering repulsion all the while, it cannot slip into an orbital of the outer shell, for all those orbitals are full. Because of the exclusion principle, it has to start a new shell and cannot approach the nucleus closely. Hence, besides being rebuffed by the electrons already present, it is only weakly attracted by the distant nucleus. The outcome is therefore an unattractive proposition for energy investors, and the doubly charged anion does not form. A similar argument applies to the other halogens. They can accept a single electron, but no more. All the *halide ions*—fluoride, chloride, bromide, and iodide—are singly negatively charged.

One step to the west from fluorine on the northern coast lies oxygen, with structure [He]$2s^2 2p^4$. This atom has two gaps in its outer shell, and perhaps its abilities are now clear. It can accept one electron readily (with the release of 1.5 eV of energy), and that electron can experience a strong attraction from the nearby central nucleus. A second electron has to be pushed onto the atom against the repulsion of the ion's negative charge, and so investment has to be made, much as in the case of the addition of a second electron to fluorine. However, now the prize is worth the effort. The second electron can find a home in the incomplete outermost shell of the singly charged oxygen anion, and settle down into a happy interaction with the not-far-off nucleus. Energy still has to be supplied to create a doubly charged oxygen anion (the *oxide ion*) from the singly charged ion, but it is of a magnitude (8.8 eV) small enough to be recovered from energy released by other steps involved in a chemical reaction. As for putting a third electron onto the oxide ion, of that there is no hope, because it would need to be forced onto an already doubly charged anion, and once it had arrived it would have to sit outside the core and would not interact appreciably with the nucleus. Thus, oxygen and the rest of its group can be expected to form doubly charged anions, which is what chemists find.

Now we can return to the largely sterile eastern shore, where the noble gases lie. All their orbitals are full, and any newly arrived electron can remain only if it occupies an orbital of a new shell. There it spreads around the atom, at a great distance from the central nucleus. There is no energy advantage in this position, and the electron affinities of the noble gases are negative. For example, to attach an extra electron to neon requires 1.2 eV, and to do the same to argon requires 1.0 eV. They are as aloof to the addition of an electron as they are zealous in retaining the electrons they have. The noble gases lie in a largely dead zone of the kingdom.

A final word about the sizes of anions. As we saw in the case of fluorine and the fluoride ion, an anion is larger than its parent atom because the additional electron puffs the atom up against the pull of the central nucleus. Superimposed on this general puffiness of anions is a variation through the triangle of the kingdom where anions are important, in the far northeast—a variation that reflects the variation in the diameters of the parent atoms. In this triangle, the landscape constructed from the diameters of anions rises from north to south and falls from west to east, and everywhere lies just a little above the landscape of the atoms themselves.

CHAPTER 11

LIAISONS AND ALLIANCES

Much, if not quite all, of the richness of the real world stems from the compounds formed by members of the kingdom. From a hundred or so elements, millions of such alliances can be constructed, just as an infinite literature can be forged from the letters of an alphabet. It would be far too demanding to wander widely in the field of all potential and actual alliances, but there are certain features showing a periodicity that matches the rhythms of the kingdom. Chemists often classify the alliances they contrive by referring to the location of the regions they have employed, and we shall do the same. Here we shall examine the broad types of alliance that may be formed, and relate the characteristics of the allied parties and the resulting compounds to locations in the kingdom.

Compounds are intimate intermarriages of the atoms of the regions, not mere mixtures. In some cases, these marriages are highly stable and persist for eons. Such is the case with the liaisons that form the Earth's core and its rocky landscapes. Other liaisons are less stable, and may be gone in a season. Such is the case with many of the naturally organic compounds based on carbon, which persist for a day or a year, or three-score years and ten, and then decay to less intricate assemblages of atoms. Some alliances are so brief as to require a highly trained eye to discern and record.

Compounds are held together by *chemical bonds*, or links between atoms. Such links are known to arise from the deployment of the electrons of the outer shells of atoms, the so-called valence shells. "Valence" is a term signifying the power of atoms to form chemical bonds, and is derived from the Latin word for "strength." *Valete!* ("Be strong!") was what the Romans said on parting. We have seen that the electronic structure of atoms displays a periodicity that is captured by the layout of the kingdom, and we can expect that the ability to form bonds will show a similar periodicity. This general periodicity of number and type of bond is what we limit our discussion to here.

There are two principal varieties of chemical bonds—*ionic bonds* and *covalent bonds*. An ionic bond, as its name suggests, is an interaction between the ions that atoms form, stemming from the attraction between the opposite charges of cations and anions. A covalent bond is a discrete combination of atoms, effected by the sharing of electron pairs. We shall examine these two types of bonds in more detail, then see how their formation correlates with an element's location in the kingdom. In each case, we shall be guided by a signpost in the kingdom that points to lower energy: bonds between atoms form if, in doing so, energy is released. We need to understand why the signpost of energy points downhill—why the resulting compound must always have a lower energy than its separate atoms.

To understand the influences leading to the formation of an ionic bond, we shall consider a gas of sodium atoms (with electronic structures [Ne]$3s^1$) mingled with chlorine atoms ([Ne]$3s^2 3p^5$), which come from the other end of the kingdom. This intermingling of atoms is still only a mixture, for the alliances have not yet been formed. We need to consider whether there is a signpost in this mixture that points downhill. There is certainly a signpost that points sharply uphill: the removal of one electron from each sodium atom, which requires 5.1 eV. Nevertheless, it may prove worthwhile to make the investment, so we make it. We have also seen that 3.6 eV of energy is released when an electron enters a chlorine atom and forms a chloride ion. At this stage, we have achieved two noble-gaslike structures, a sodium cation and a chlorine anion, by the transfer of an electron from one type of atom to the other. The signpost still points uphill, though, because 5.1 eV minus 3.6 eV equals 1.5 eV, and there seems to be no reason why this gas of ions should form.

However, there is an all-important third contribution: the electric attraction between the ions of opposite charge. If the ions approach one another, this attraction may become so strong that the lowering of energy outweighs the investment we have had to make, and as soon as the interaction energy exceeds 1.5 eV, the signpost swings like a semaphore and points downhill. The lowest energy of all is reached if the cations and anions clump together, with cations around each anion and anions around each cation. More detailed determinations of the resulting structure, which is called an *ionic solid*, show that there are six cations around each anion and six anions around each cation (fig. 14). This is in fact the structure of common table salt, the sodium chloride won from mines and oceans.

We shall use this discussion to examine the types of compounds formed by the elements and expose more of the rhythms of nature that the kingdom displays. These

FIG. 14.

The structure of common salt (sodium chloride): an exploded view on the left and the actual structure on the right. This pattern continues potentially indefinitely, and comes to an end only at the edge of a crystal. Note that each sodium cation (the small spheres) is in contact with six chloride ions (the large spheres), and vice versa.

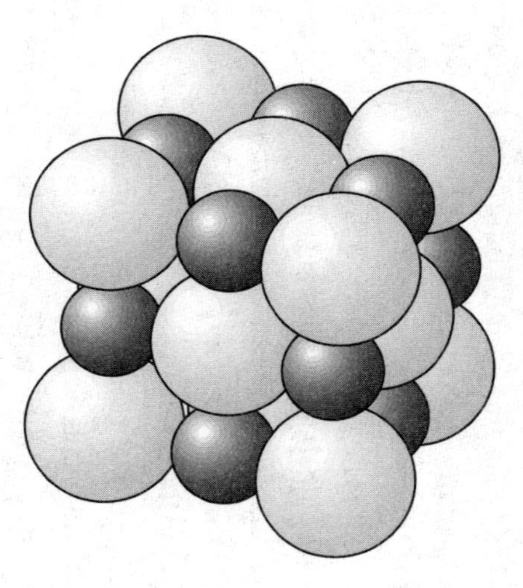

rhythms of chemical combination were in fact among the most compelling components of the evidence that led Dmitri Mendeleev to his map. First, note that sodium can release only one electron readily, and that, similarly, chlorine can accept only one. The most plausible combination of these two distant elements is one sodium atom and one chlorine atom, which is exactly the proportions in which they combine to form salt. Indeed, we can use our command of the electronic geography of the kingdom to predict that *every* compound formed from an alkali metal from Group 1 with a halogen from Group 17 will have atoms present in equal proportions. That is observed, and all alkali metal halides, as they are called (including lithium fluoride, sodium bromide, potassium iodide, and all thirty binary combinations that are possible for these two zones) have this one-to-one composition. Now the laws of the kingdom, the underlying electronic structures, are governing alliances, too.

Now suppose that we move east one group, to calcium and the alkaline earth metals of the Western Desert. These elements can readily surrender two electrons before exposing their inviolable cores. The halogens, though, can still only accept one electron per atom. This suggests that when the compounds form, there will be two halogen atoms for each alkaline-earth-metal atom. That is exactly the composition that is found.

We can now reverse the argument. Suppose we return to the alkali metals of Group 1 but now consider the alliances that they may form with the regions of Group 16, oxygen and its southerly cousins. The former can surrender one electron each; the atoms of the latter can each accept two. So now we expect compounds in which there are two alkali-metal atoms for each oxygen or sulfur atom, exactly as is found.

To round off the discussion, we now move simultaneously east into Group 2 and west into Group 16. The alkaline-earth-metal atoms can surrender two electrons, and the oxygenlike atoms can each accept two. So once again we have the pattern of compounds in which for each atom of Group 2 there will be one atom of Group 16. This pattern matches exactly what is observed, and accounts for the composition of compounds like lime (calcium oxide, one calcium ion to one oxide ion).

This kind of argument is used by chemists whenever they want to explain or predict the composition of an ionic solid. The overall signpost of energy points downhill only if the amount of energy required to form cations is not too large: a too-large investment in cation formation may never be recovered from the attraction between opposite charges, and the signpost will never point downhill. That in turn limits the formation of ionic bonds to combinations of elements in which one is a metal, for only metallic elements have sufficiently low ionization energies. Thus, when we view the kingdom from above and see the Western Desert glittering below, we now know that these are the regions that have the potential to form ionic solids, and that the compositions of the solids they form are dictated by the location of their group in the kingdom.

Ionic solids have a variety of features in common, which make them quite easy to recognize in the real world. First, because they are tough aggregates of ions stacked tightly together, they are rigid, brittle solids. To turn them to liquids, it is necessary to heat them to a temperature high enough to shake the ions loose from one another (heat shakes things), so ionic solids usually have high melting points. A further important characteristic is that when they dissolve in water (though by no means all of them do), the ions drift apart and become mobile conductors of electricity. Thus, ionic solids are potential *electrolytes*, or substances that can carry an electric current either when molten or when dissolved.

Now we consider the formation of covalent bonds. When a compound is formed that does not involve a metallic element, too much energy is involved in electron extraction to make an ionic bond a feasible proposition. The signpost points uphill in energy, even after allowing for the attraction between the ions that may ultimately be produced. The best that can happen is for the atoms effectively to retain their electrons but to enter into a sharing agreement. When two electrons are shared between two neighboring atoms, we say that they are linked by a covalent bond.

First, we need to see why two electrons constitute a bond, and not one electron, or three electrons, or some other number. The reason can be traced back to the Pauli exclusion principle, which limits the occupation of atomic orbitals to no more than two electrons. When two atoms come together, the distribution of the electrons of their valence shells is no longer confined to each atom alone but now spreads over both of them like a net. Analogous to the distributions in atoms—the atomic orbitals—the distributions in molecules, which spread over the constituent atoms, are called *molecular orbitals*. However, even though they have a wider span and a more complicated form, these orbitals are still orbitals, and the exclusion principle still applies: only two electrons can be accommodated in one of these orbitals, and this is why a covalent bond consists of a pair of electrons.

Incidentally, it is also possible for two atoms to share more than two electrons, because more than one molecular orbital can spread over them and net them together. Each shared pair of electrons counts as a covalent bond, and so atoms may be bound by single bonds (one shared pair), double bonds (two shared pairs), triple bonds, and—very, very rarely—quadruple bonds. As in the formation of ionic bonds, covalent bonds will form only if they result in a lowering of energy; but in the case of covalent bonds we do not have to worry about the huge investment that ion formation generally demands, for electron sharing involves much less drastic redistributions of electrons. Covalent bonds are typically formed between elements that lie in the upper triangle of the Eastern Rectangle.

When covalent bonds form between atoms, the resulting entity is called a *molecule*. It is here that the compounds forged by the formation of covalent bonds differ so strikingly from those formed by ionic bonds. Molecular compounds are typically (but not invariably, as we shall see) discrete groupings of atoms rather than extended aggregations. Moreover, because electron sharing is more subtle than electron loss, the ability of an atom partially to release electrons to form a covalent bond in one direction will affect its ability to release them in a different direction. As a result, the arrangement of atoms in a molecule has a fixed, characteristic geometry. That is, the covalent compounds of the kingdom are discrete, often small groupings of atoms—groupings with characteristic shapes.

Whereas ionic aggregations are potentially infinite and invariably solid at normal temperatures, molecular aggregations are often so small that they can form gases and liquids. Moreover, when they form solids, the forces of attraction between the molecules in the solid are generally much less strong than between ions in an ionic solid. Many molecular species form softer solids than ions do, and are more easily shaken apart into their constituent molecules by the gentle application of heat; they are therefore commonly characterized by low melting points and low boiling points.

In general, molecular compounds are the soft face of nature, and ionic compounds are the hard. Few distinctions make this clearer than those between the soft face of the Earth—its rivers, its air, its grass, its forests, all of which are molecular—and the harsh substructures of the landscape, which are largely ionic. This is why the upper triangle of the Eastern Rectangle is so important to the existence of life, and why all the rest of the kingdom is so important in the formation of a stable, solid platform.

It is not invariably the case that covalent bonds result in soft alliances. There are certain cases in which atoms can form covalent bonds to neighbors, those neighbors can bind neighbors, and so on, to form a potentially infinite solid. One example is diamond, a form of carbon. The immense hardness of this form of the element almost makes up for its contribution to softness elsewhere, and stems from the rigid latticelike bonding that extends throughout the solid and binds atom to atom virtually without end, like the steel frame of a large building.

There are several points we need to tie up before leaving this aspect of alliances in the kingdom. One concerns the overarching power of carbon to participate in molecule formation, a power that results in such complexity of structure and collaboration that the alliances it forms become alive and can reflect upon themselves. The essential reason for this latent power of a single region is, as noted, carbon's intrinsic mediocrity, its lack of self-assertion. Sitting as it does in the middle of the northern coast, it is neither an aggressive shedder of electrons, as are elements to the left, nor is it an avid receiver, like the atoms to its right. Carbon is mild in its demands on the alliances it makes. Moreover, it is even content with its own company, and can make extensive liaisons with itself, forming chains, rings, and trees of atoms. Were it readier to give up its electrons, it would do so at the demand of another atom, and then find that it had not retained sufficient electrons to bind other atoms in a precise disposition. If it were more avid for electrons, it would soon satisfy its tendency to bond and would lack the opportunity for subtle conspiracy with others. By being in the middle, undemanding and not particularly generous, it can spin lasting alliances rather than hasty conspiracies.

A final point concerns the inert eastern edge of the kingdom, the noble gases at the seashore of nonreactivity. They are largely dead to the world of bond formation, even covalent-bond formation, for reasons that can be traced to their electronic structures. Their ionization energies are high, because their electrons form a compact clump around their highly charged nuclei. They are reluctant to release electrons even partially, for that, too, demands a substantial investment of energy. Nor do these closed-shell species attract electrons, for any arriving electron faces the prospect of inhabiting an orbital of a new shell, undesirably far from the nucleus. Here the signpost of energy hardly ever points down, and alliances are hardly ever formed, except with certain politically aggressive regions, such as fluorine. That the kingdom plunges to the shore abruptly at its eastern edge is a sign of satiation. Why trouble to acquire more electrons? Why trouble to give any up?

EPILOG

We have traveled widely in the kingdom, surveying it from high in the air in different lights, and trekking through it on the ground, where we have seen its texture. We have burrowed beneath its surface and seen the inner workings of its laws. Now, at the end of this journey, we should collect our thoughts on the subject of this imaginary land and its importance.

The real world is a jumble of awesome complexity and immeasureable charm. Even the inanimate, inorganic world of rocks and stone, rivers and ocean, air and wind is a boundless wonder. Add to that the ingredient of life, and the wonder is multiplied almost beyond imagination. Yet all this wonder springs from about one hundred components that are strung together, mixed, compacted, and linked, as letters are linked to form a literature. It was a great achievement of the early chemists—with the crude experimental techniques available but also with the ever-astonishing power of human reason (as potent then as now)—to discover this reduction of the world to its components, the chemical elements. Such reduction does not destroy its charm but adds understanding to sensation, and this understanding only deepens our delight.

Then there came a grander achievement. Although these elements are matter, and few thought that one might be related to another, chemists saw through the superficialities of appearance and identified a kingdom of relationships, of family ties, of alliances, of affinities. Through their experimentation and reflection, a land rose from the sea, and then the elements were seen to form a landscape. Just as important, the landscape, when examined in a wide variety of lights, was seen to be structured, not merely a random collection of gorges and pinnacles; in particular, it was found to be periodic in variation. That was the most astonishing discovery of all, for why should matter display any sort of periodicity?

As so often in the development of science, comprehension springs from simple concepts that operate just below the surface of actuality, and constitute the true actuality. Once atoms were known—and their constitution elucidated in terms of that great invention of the mind, quantum mechanics—the foundation of the kingdom was exposed. Simple principles—the enigmatic exclusion principle, in particular—showed that the periodicity of the kingdom was a representation of the periodicity of the electronic structure of atoms.

The structure, layout, and probable extensions of the kingdom are now fully understood. There are deeper currents flowing than we have presented here, but those currents are known. Yet despite our comprehension of the kingdom, it remains a mysterious place. The properties of the regions are rationalizable, and within certain ranges we can predict with confidence the chemical and physical properties of an element and the types of compounds it forms. The kingdom—the periodic table—is the single most important unifying principle of chemistry: it hangs on walls throughout the world, and the best way of mastering chemistry and inspiring new lines of chemical research lies in an understanding and use of its layout. But it is a land of conflict, as we have seen. The properties of a particular region are the outcome of competition, of influences that pull in different directions, sometimes several influences. These influences are usually finely balanced, and it is difficult, even from the heights of experience, to be absolutely sure that an element will not have a particular quirk of personality that will defeat a prediction or open up a new and exciting avenue of investigation.

Just as the letters of the alphabet have the potential for endless surprise and enchantment, so, too, do the elements of the kingdom. Unlike an alphabet, which has hardly any infrastructure, the kingdom has sufficient structure to make it an intellectually satisfying aggregation of entities. And because these entities are finely balanced, living personalities, with quirks of character and not always evident dispositions, the kingdom will always be a land of infinite delight.

FURTHER READING

Atkins, P. W., *Quanta: A Handbook of Concepts*, 2d ed. (Oxford: Oxford University Press, 1991). The book provides short descriptions of the quantum mechanical concepts encountered in this book.

Atkins, P. W., and J. A. Beran, *General Chemistry*, 2d ed. (New York: Scientific American Books, 1992). This volume introduces the principal concepts described in this book at an elementary level, including modern atomic structure and examples of periodic chemical character.

Bourne, J., "An Application-Oriented Periodic Table of the Elements," *Journal of Chemical Education*, 66 (1989): 741–45. Several different versions of the table are presented.

Cox, P. A., *The Elements: Their Origin, Abundance, and Distribution* (Oxford: Oxford University Press, 1989). An introduction to the processes that formed elements and distributed them through the cosmos and the Earth.

Emsley, J., *The Elements*, 2d ed. (Oxford: Oxford University Press, 1991). A collection of numerical data on the elements in a convenient format.

Mason, J., "Periodic Contractions Among the Elements: Or, on Being the Right Size," *Journal of Chemical Education*, 65 (1988): 17–20. A survey of atomic and ionic sizes, their periodicity, and the consequences.

Mazurs, E. G., *Graphical Representations of the Periodic System During One Hundred Years* (Alabama: University of Alabama Press, 1974).

Puddephatt, R. J., and P. K. Monaghan, *The Periodic Table of the Elements*, 2d ed. (Oxford: Oxford University Press, 1986). An introductory survey of periodic trends.

Ringnes, V., "Origin of the Names of Chemical Elements," *Journal of Chemical Education*, 66 (1989): 731–38. A thorough and interesting account of the names of the elements.

Rouvray, D. H., "Turning the Tables on Mendeleev," *Chemistry in Britain* (May 1994): 373–78. A short survey of the history of the periodic table.

van Spronsen, J. W., *The Periodic System of Chemical Elements: A History of the First Hundred Years* (Amsterdam: Elsevier, 1969).

Weast, R. C., ed., *CRC Handbook of Chemistry and Physics*, 76th ed. (Boca Raton: CRC Press, 1995). The handbook includes brief descriptions of the history of the discovery of the elements.

Weinberg, S., *The First Three Minutes* (London: Andrew Deutsch, 1977). A classic popular account of the formation of matter.

Woods, G., "The Deeper Picture," *Chemistry in Britain* (May 1994): 382–83. A brief account of various alternative arrangements of the periodic table that have been proposed.

INDEX

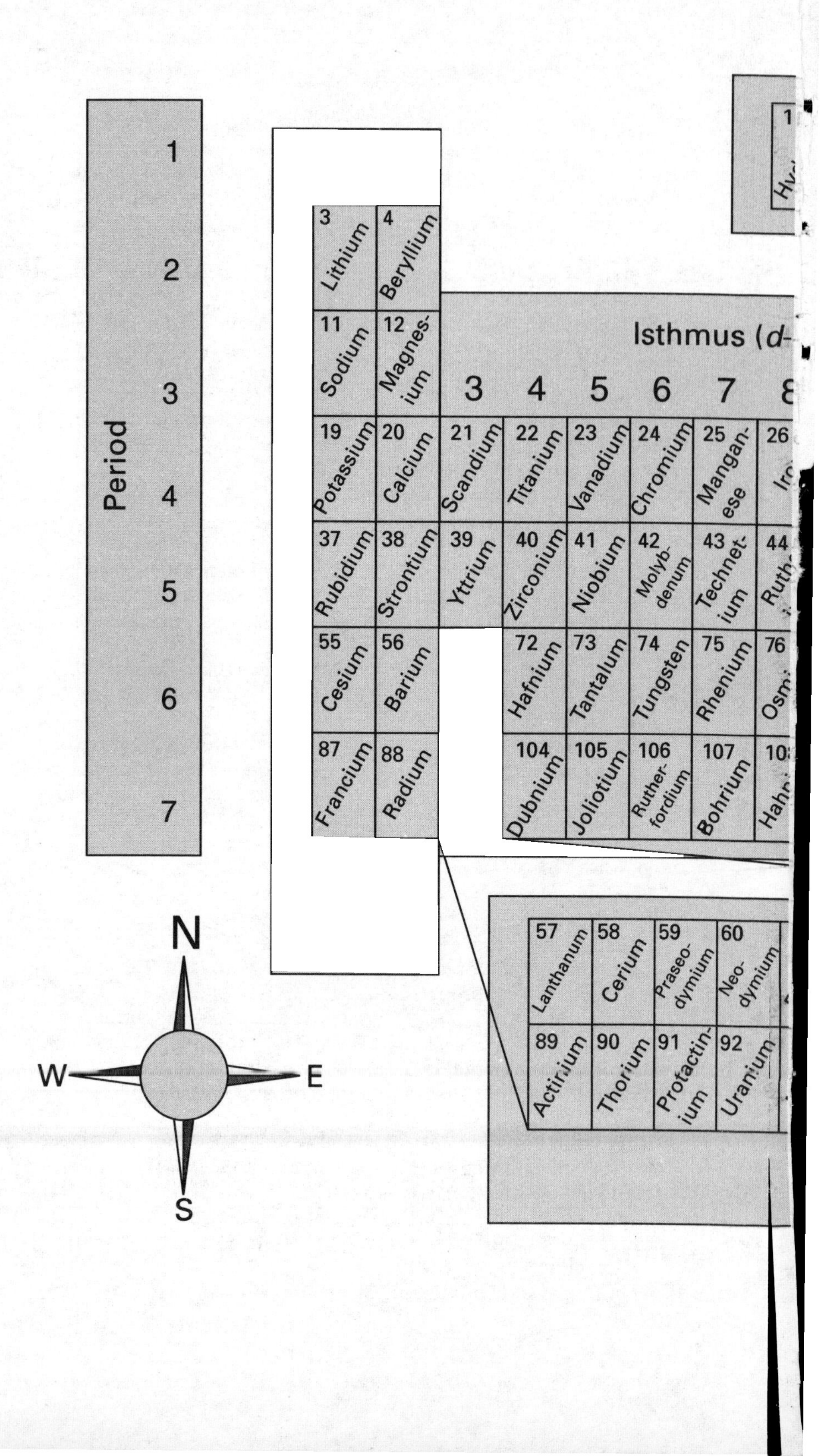

Period
1
2
3
4
5
6
7
3 Lithium
4 Beryllium
11 Sodium
12 Magnes-ium
19 Potassium
20 Calcium
37 Rubidium
38 Strontium
55 Cesium
56 Barium
87 Francium
88 Radium
Isthmus (d-
3
4
5
6
7
21 Scandium
22 Titanium
23 Vanadium
24 Chromium
25 Mangan-ese
26
39 Yttrium
40 Zirconium
41 Niobium
42 Molyb-denum
43 Technet-ium
44
72 Hafnium
73 Tantalum
74 Tungsten
75 Rhenium
76
104 Dubnium
105 Joliotium
106 Ruther-fordium
107 Bohrium
57 Lanthanum
58 Cerium
59 Praseo-dymium
60 Neo-dymium
89 Actinium
90 Thorium
91 Protactin-ium
92 Uranium
N
W
E
S

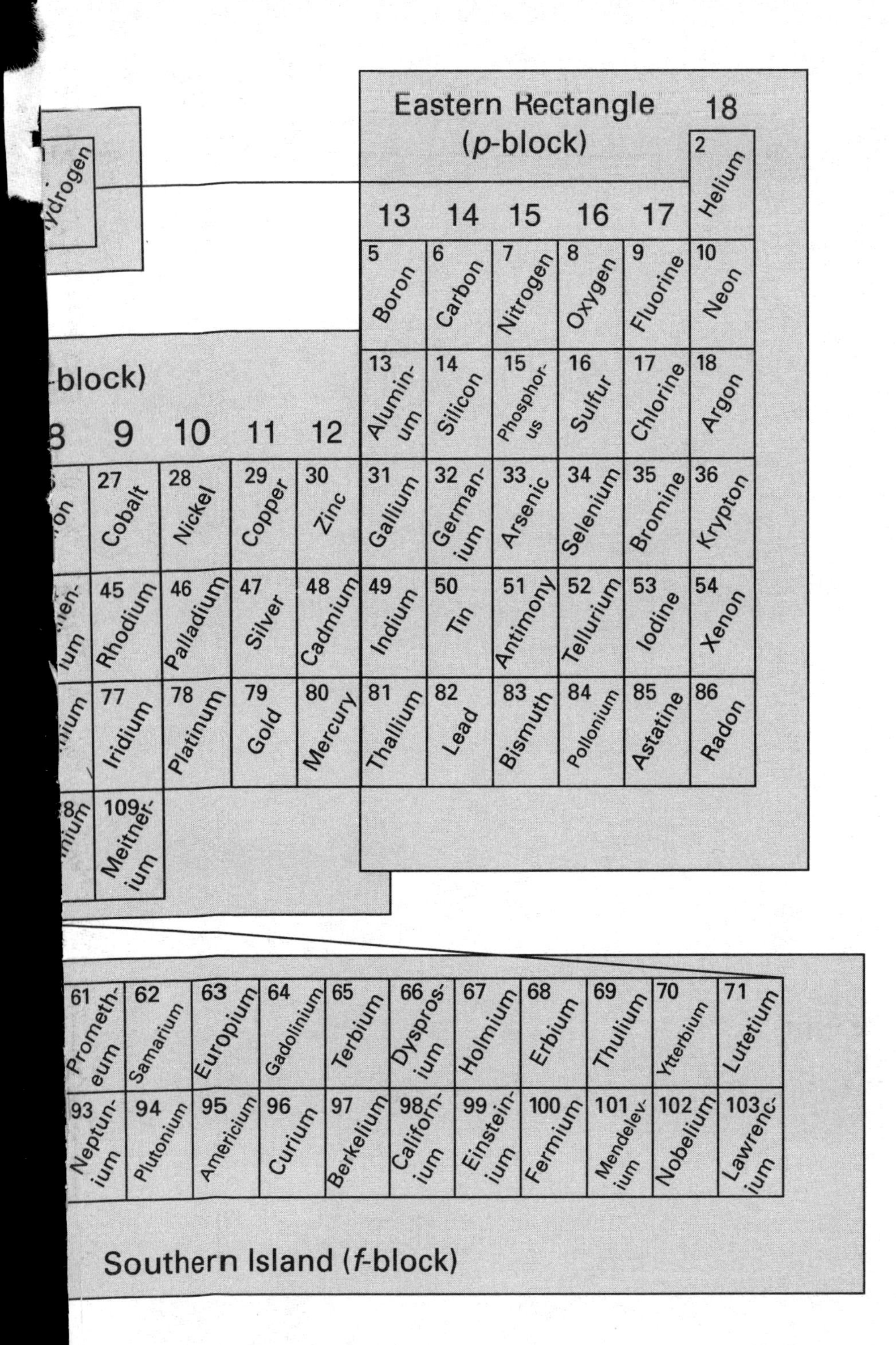

Eastern Rectangle (p-block)
18
2 Helium
13 14 15 16 17
5 Boron
6 Carbon
7 Nitrogen
8 Oxygen
9 Fluorine
10 Neon
13 Aluminum
14 Silicon
15 Phosphorus
16 Sulfur
17 Chlorine
18 Argon
31 Gallium
32 Germanium
33 Arsenic
34 Selenium
35 Bromine
36 Krypton
49 Indium
50 Tin
51 Antimony
52 Tellurium
53 Iodine
54 Xenon
81 Thallium
82 Lead
83 Bismuth
84 Pollonium
85 Astatine
86 Radon
-block)
9 10 11 12
27 Cobalt
28 Nickel
29 Copper
30 Zinc
45 Rhodium
46 Palladium
47 Silver
48 Cadmium
77 Iridium
78 Platinum
79 Gold
80 Mercury
109 Meitnerium
61 Promethеum
62 Samarium
63 Europium
64 Gadolinium
65 Terbium
66 Dysprosium
67 Holmium
68 Erbium
69 Thulium
70 Ytterbium
71 Lutetium
93 Neptunium
94 Plutonium
95 Americium
96 Curium
97 Berkelium
98 Californium
99 Einsteinium
100 Fermium
101 Mendelevium
102 Nobelium
103 Lawrencium
Southern Island (f-block)